Design Against Fire

Design Against Fire

An introduction to fire safety engineering design

Edited by

Paul Stollard
Rosborough Stollard Ltd

and

Lawrence Johnston
Queen's University of Belfast

E & FN SPON
An Imprint of Chapman & Hall
London · Glasgow · New York · Tokyo · Melbourne · Madras

Published by E & FN Spon, an imprint of Chapman & Hall, 2–6 Boundary Row, London SE1 8HN, UK

Chapman & Hall, 2–6 Boundary Row, London SE1 8HN, UK

Blackie Academic & Professional, Wester Cleddens Road, Bishopbriggs, Glasgow G64 2NZ, UK

Chapman & Hall Inc., One Penn Plaza, 41st Floor, New York NY 10119, USA

Chapman & Hall Japan, Thomson Publishing Japan, Hirakawacho Nemoto Building, 6F, 1–7–11 Hirakawa-Cho, Chiyoda-ku, Tokyo 102, Japan

Chapman & Hall Australia, Thomas Nelson Australia, 102 Dodds Street, South Melbourne, Victoria 3205, Australia

Chapman & Hall India, R. Seshadri, 32 Second Main Road, CIT East, Madras 600 035, India

First edition 1994

© 1994 Paul Stollard and Lawrence Johnston

Typeset in 10/12pt Palatino by Type Study, Scarborough
Printed in Great Britain by T. J. Press Ltd, Padstow, Cornwall

ISBN 0 419 18170 9

Apart from any fair dealing for the purposes of research or private study, or criticism or review, as permitted under the UK Copyright Designs and Patents Act, 1988, this publication may not be reproduced, stored, or transmitted, in any form or by any means, without the prior permission in writing of the publishers, or in the case of reprographic reproduction only in accordance with the terms of the licences issued by the Copyright Licensing Agency in the UK, or in accordance with the terms of licences issued by the appropriate Reproduction Rights Organization outside the UK. Enquiries concerning reproduction outside the terms stated here should be sent to the publishers at the London address printed on this page.

The publisher makes no representation, express or implied, with regard to the accuracy of the information contained in this book and cannot accept any legal responsibility or liability for any errors or omissions that may be made.

A catalogue record for this book is available from the British Library

Library of Congress Cataloging-in-Publication data

Design against fire : an introduction to fire safety engineering design / edited by Paul Stollard and Lawrence Johnston. – 1st ed.
   p.   cm.
Includes index.
ISBN 0–419–18170–9 (alk. paper)
  1. Building, Fireproof. 2. Fire prevention – Equipment and supplies. I. Stollard, P. (Paul), 1956– . II. Johnston, Lawrence.
TH1065.D47 1993
693'.82–dc20   93–32971
                                                                                 CIP

∞ Printed on permanent acid-free text paper, manufactured in accordance with ANSI/NISO Z39.48–1992 and ANSI NISO Z39.48–1984 (Permanence of Paper).

Contents

Notes on contributors	ix
Preface	xi
Introduction: 'Together Against Fire'	1
Jack Anderson	

1 Fire science — 9
Dougal Drysdale
  1.1 Introduction — 9
  1.2 The critical stages of fire — 9
  1.3 Prevention of ignition — 10
  1.4 Prevention or delay of flashover — 11
  1.5 Life safety — 14
  1.6 Property protection — 16
  1.7 Application of fire science to fire safety — 18

2 Fire safety engineering — 21
Paul Stollard
  2.1 Introduction — 21
  2.2 Tactics — 22
  2.3 Components — 23
  2.4 Acceptability and equivalency — 27
  2.5 Traditional and holistic approaches to fire safety design — 29

3 Fire prevention – designing against arson attack — 32
Lawrence Johnston
  3.1 Introduction — 32
  3.2 Three lines of defence — 33
  3.3 The briefing, design and construction process — 35
  3.4 Types of arson — 37
  3.5 Balancing risks — 38

4 Fire communications — 41
John Northey
  4.1 Introduction — 41

4.2	Choosing the detector	41
4.3	Heat detectors	43
4.4	Smoke detectors	45
4.5	Sampling/aspirating detectors	45
4.6	Flame detectors	46
4.7	Manual call points	46
4.8	Warning devices	47
4.9	Control equipment and systems	48
4.10	Installation requirements	50
4.11	Indicating the fire	52
4.12	Users' responsibilities	53

5 Escape behaviour in fires and evacuations — 56
Jonathan Sime

5.1	Introduction	56
5.2	Design × Information Technology × Management × Buildings in Use	57
5.3	Models of human movement and behaviour in emergencies	59
5.4	Research on escape behaviour	60
5.5	Exit choice behaviour	63
5.6	Escape behaviour factors, assumptions and principles	64
5.7	Research study examples	66
5.8	Implications of research for fire safety engineering design	79

6 Fire escape in difficult circumstances — 88
John Abrahams

6.1	Introduction	88
6.2	Occupancy characteristics	89
6.3	Escape strategies	92
6.4	A 'first principles' approach to escape	94

7 Principles of fire containment — 97
H.L. Malhotra

7.1	Introduction	97
7.2	Historical aspects	99
7.3	Fire compartmentation needs	100
7.4	Essential compartmentation	103
7.5	Optional compartmentation	105
7.6	Fire resistance requirements	105
7.7	Fire resistance provision	107
7.8	Available technologies	108
7.9	Future needs	109

Contents vii

8 Smoke control in shopping malls and atria   **110**
Howard Morgan
    8.1 Introduction   110
    8.2 A brief history of smoke ventilation   111
    8.3 Basic principles of smoke ventilation   112
    8.4 Design parameters for atria and malls   114
    8.5 Activation of the system   119

9 Fire information   **121**
Paul Stollard
    9.1 Introduction   121
    9.2 Legislation in Great Britain   122
    9.3 Northern Ireland legislation   131
    9.4 British Standards and international standards   132
    9.5 Guidance   138
    9.6 Consultancy and advisory services   149

Appendix: Details of the 'Design Against Fire' course   153
Glossary   161
Index   167

Notes on contributors

John Abrahams is estate safety officer for the South Western Regional Health Authority. He has been a consultant to Rosborough Stollard Ltd, architectural technologists, since its establishment and is co-author of the basic undergraduate text, *Fire from First Principles*. He is an external tutor for the College of Estate Management, Reading.

Jack Anderson has a background of practice in Scotland and London, and in teaching and research in the UK and USA. He has chaired the RIBA Research Group for the past five years and was recently appointed Chairman of the Building Regulations Advisory Committee at the Department of the Enviroment. He has been a partner in Bickerdike Allen since 1980.

Dougal Drysdale is Reader and Director of the Unit of Fire Safety Engineering at the University of Edinburgh. His research interests lie in fire science, fire dynamics and the fire behaviour of combustible materials.

Lawrence Johnston is a lecturer in the Department of Architecture and Planning at the Queen's University of Belfast and jointly directed the 'Design Against Fire' course. He is also a partner in Leighton Johnston Associates (Architects), based in Belfast.

H. L. Malhotra (Bill) after spending most of his working life at the Fire Research Station has been propagating the philosophy and practice of fire engineering as director of AGNICONSULT, a consultancy service specializing in finding engineering solutions to fire safety problems.

Howard Morgan has a degree in Physics from the University of Manchester and joined the Fire Research Station in 1973. His research has included small-scale buildings, and ways of controlling heat movement. More recently his section's work has involved the study of the interactions between different active fire protection techniques (e.g. between venting and sprinklers).

John Northey is the editor of *Fire Surveyor* and Chairman of the European Committee on automatic fire detection and fire alarm systems, and also Chairman of British Standards technical sub-committee FSM12/1 which is responsible for UK fire alarm and detection systems.

Notes on contributors

Jonathan Sime is an environmental psychologist with 20 years' experience in research, lecturing and consultancy, with particular reference to human behaviour in fires and crowd safety, a focus for his publications. He is secretary of the International Association for People–Environment Studies (IAPS) and a consultant to various bodies such as the Channel Tunnel Safety Authority.

Paul Stollard is a director of Rosborough Stollard Ltd, architectural technologists. He has been involved in fire safety education, research and design for the past ten years and is co-author of the basic undergraduate text, *Fire from First Principles*. He has been Visiting Professor in Architecture at the Queen's University of Belfast, and while there he jointly directed the Design Against Fire course.

Illustrations in the text were prepared by Derrick Perkins. The two illustrations in Appendix A were prepared by Bronagh Long and Martina Ennis.

Preface

Traditionally architects have regarded Fire Prevention Officers and Building Control Officers as if they were the construction industry's equivalent of traffic wardens. While an architect will admit that they are necessary to prevent other people behaving badly and putting up dangerous or unsafe buildings, the same architect will become defensive and annoyed when these statutory authorities begin to take a particular interest in his own projects. All too often, building regulations and fire safety standards have been seen as obstacles, rather than aids, to good design. Architects may talk about 'what they can get away with', or what is 'the minimum they need to do to comply'. This confrontational approach has not been helpful in the past, especially when dealing with new or innovative buildings. With the more recent moves towards functional requirements, rather than prescriptive building regulations, and the growth in specifically fire engineering solutions, such an attitude is positively dangerous.

The Bickerdike Allen Report, *Fire and Building Regulation* (HMSO, 1990) into the working of the fire legislation in England and Wales recognized the need for architects and statutory authorities involved in fire safety to work closely together rather than in opposition. One of the Report's recommendations was that there should be courses to train all three groups (architects, Building Control Officers and Fire Prevention Officers) in the basics of fire engineering design. The Report also advocated that such training would be even more valuable if it were done jointly, so that prejudices and misconceptions could be avoided.

In an attempt to follow these recommendations, the various relevant groups in Northern Ireland worked together to establish joint professional development seminars. These were not designed for inexperienced students, but for professionals already experienced in their own fields. The first of the courses was run in the academic year 1991/92, and they are now held regularly. They are unique in being both completely multi-disciplinary and providing a postgraduate certificate recognized by a university. Although organized by the Queen's University of Belfast, there are also significant contributions from both the Fire

Authority for Northern Ireland and the six building control groups in the province.

The course members learn not only directly from the lectures, projects and visits, but most important, from each other. Northern Ireland was ideally suited to launch such an initiative as it is served by a single fire authority, and a coherent grouping of building control authorities. It also has its own legislative system, with distinct Building Regulations and Fire Prevention Orders. Most development within Northern Ireland is designed by local practices, and fostering good relations between these and the authorities was a key objective behind this initiative.

As this was an innovative course, it was possible to attract the very best experts to give what in effect were 'master classes'; these included Jack Anderson, Dougal Drysdale, John Northey, Jonathan Sime, John Abrahams, Bill Malhotra and Howard Morgan. They were supported by detailed discussion of the relevant legislation provided by the local statutory authorities. Then working as multi-disciplinary teams, the course members were confronted with design problems to solve.

When it came to fire-fighting and fire extinguishment, those on the course were given the opportunity to deal with a real fire at the Fire Brigade's training school. Having experienced together the difficulties of fighting even a training fire, in thick smoke and wearing breathing apparatus, the course members found it somewhat easier to work together on the design projects.

This book brings together, in the first eight chapters, the keynote papers presented on the first course in a form which, it is hoped, will make them accessible to architects, Fire Brigade Officers and Building Control Officers. Written by specialists from different fields, the chapters do not appear in a consistent form or style, instead they reflect the varying subjects under consideration. Some chapters are scientific or research based, with references and case studies; some have been written by estate professionals with responsibility for fire safety and are simple descriptions of the different problems faced; and others are more concerned with the legislation and codes of practice. This variety reflects the different skills needed by the fire engineer, who has to be able to combine scientific principles, research methodology, design experience and an understanding of the legislation.

In the last chapter (Chapter 9) it is intended to provide a summary of all the more detailed sources of information available to the design team. The basic elements of fire safety legislation, both in Northern Ireland and in the rest of the British Isles, are outlined. The key bodies and additional sources of advice are listed and there is a glossary of relevant technical terms.

There is a risk of fire in every building that is designed, and it therefore has to be accepted that complete safety from fire is an impossible goal.

However, it is morally right that what can be done to reduce that risk *is* done. Fire safety is not the only objective which the architect in designing a new building fulfils. There are a whole series of other objectives (aesthetic, functional, technological and economic) which must also be satisfied, and if the design is to be successful, these must be integrated into a coherent whole during the design process. It is the architect's responsibility to ensure that the objectives of fire safety are integrated with the more general objectives of architectural design. It is hoped that this series of papers will outline some of the fundamental principles of fire safety, so that the architects, Fire Prevention Officers and Building Control Officers can work as a team to ensure a safe and successful design.

Introduction: 'Together against Fire'

Jack Anderson

The past three years have seen much activity and some changes in the field of fire safety in buildings. One way or another, as architects and fire consultants, Bickerdike Allen Partners (BAP) have been much involved; and from that experience, this Introduction will touch upon some of these developments and offer some views on the events, the trends they represent and the thinking that lies behind the trends. It must be made clear that although the author is, at the time of writing, the Chairman of the Building Regulations Advisory Committee (BRAC), any views or comments expressed here are personal and those of an architect in practice.

The Department of Trade and Industry's (DTI) Enterprise and Deregulation Unit commissioned BAP in 1989 to review ways in which fire precautions and Building Regulations work in the UK and to recommend any improvement thought necessary. The first objective was to find out how the system was viewed by those seeking statutory approval of building control and fire authorities and to see what problems they perceived. Extensive canvassing of opinion revealed only moderate dissatisfaction on their part, arising not from their direct contact with the authorities, but from perceived difficulties between the two authorities. As a result, the study concentrated on ways of reducing friction within the system.

The resulting Report was published in 1990 [1] and the two main thrusts of the recommendations were procedural and educational. The procedural issues arose mainly from the idea that there is a long-term trend towards a body of legislation and regulation which is increasingly comprehensive but which, at the same time, leaves more and more to the interpretation and judgement of all participants, namely the applicants for approval, the building control authority and the fire authority. So some of the recommendations dealt with guidance through the consultative process, trying to set out a path along which each of the players in the course of achieving fire safety in new building would know not only his own role at various points along the way, but also the roles of the others involved. These included a guidance document; an arrangement for the earlier identification and, where necessary, determining of irreconcilable differences; simpler fire compliance drawings; a Part B Compliance Certificate; and the amendment of local Acts where necessary to avoid difficulties.

In some cases, the law requires formal consultation between the building control authority and the fire authority, but of much greater significance is the informal consultations that take place between these authorities and applicants at various times during the design and construction of a building. From the pattern of response, problems with this sort of consultation are most likely to occur where the building is large and its design development long, but friction and feelings aroused by the few larger applications can sour the general working of the system. Also where previously these large, complex building projects were few and metropolitan, today they are more numerous and widespread, affecting more applicants and authorities than before.

To illustrate this difficulty, the Report describes a possible but fortunately rare scenario in which the Fire Prevention Officer gives advice on a design at one point, the Building Control Officer subsequently approves a much modified design and, later still, a completed building to a further changed design is inspected for a Fire Certificate. Clearly in such a process there is a lot of room for misunderstanding and friction, but even so it would be wrong to try to prevent the fullest consultation between the three participants in the process. So ways have to be found to overcome or avoid the difficulties.

Clarifying the process itself was one obvious necessity and several of the recommendations put forward steps to achieve that. But at the same time, discussions with representatives of the building applicant, fire officers and building authorities identified the need to look at their different backgrounds as a source of problems with consultation. As set out in the report, the main participants in the process fall into three broad categories: those who design and build; those who regulate building on behalf of the public; and those whose chief concern is fire, including fire in buildings. Each category is made up of several levels of people such as the building owner, developer, builder and architect, the Building Control Officer and district council, the Fire Prevention Officer, Fire Brigade and fire authority. For simplicity, the report used the following general terms and definitions:

The Applicant Any person or group who applies for approval to build or adapt buildings, including owners, developers, architects or builders.

Introduction: 'Together Against Fire'

The Building Control Officer (BCO)	The Building Control Officer of the authority responsible for the enforcement of the Building Regulations.
The Fire Prevention Officer (FPO)	The Fire Prevention Officer representing the fire service and fire authority, responsible for advising on fire safety in general and for enforcing the Fire Precautions Act 1971.

The applicant sets the process in motion when he first signals an intention to build or alter a building. At that stage, his professional team should already have, or be able to acquire, expert advice on fire safety consistent with the complexity of the design proposed or intended. In the Fire Research lecture 'Fire and the architect' [2], Dr William Allen refers to the process 'by which architects, uninterested or ignorant about fire in buildings, negotiate with fire officers "with a mixture of reluctance and gamesmanship" and learn most of what they know about fire "by osmosis in arguments with fire officers"'.

It is assumed that it will usually be the Fire Officer, not the Building Control Officer or the private fire consultant, with whom this ill-equipped agent to the applicant will interact in seeking a safe and sensible solution to fire in design, and the Fire Officer's willingness to respond to every approach and even to take initiatives in offering advice to applicants dampens the tendency for architects to educate themselves properly in the best means of integrating fire safety measures into their building designs.

So the general level of knowledge and competence of many applicants and their agents in respect of fire development, spread and protection remains low; Building Control Officers are critical of applicants who submit substantial building projects for compliance with fire regulations in the Building Regulations without a grasp of general concepts, sufficient knowledge or specialist advice on the subject. Similarly, although in the past Fire Brigades have given free advice in such circumstances under the Fire Services Act 1947, some fire authorities now consider that this is being abused and that there should be a charge for such a service.

The training of Building Control Officers was originally structured to suit the local government system, but since the Building Act 1984 people trained as Building Control Officers may be employed by the private sector and operate in competition with their local authority counterparts. This development has brought about the need to review established training programmes to cater for the development roles of private and local authority Building Control Officers.

Entry to the profession is based on polytechnic or equivalent standards, and the profession is attracting a growing number of graduates from other disciplines which should have an enriching and broadening effect and provide the student with material capable of taking the study of special areas such as fire in buildings to postgraduate levels. The educational background of Fire Prevention Officers is not generally as high as that of BCOs. Their curriculum comprises a series of short periods of intensive training on an extremely wide and ambitious syllabus which raises some serious doubts about the depth to which any aspect can be taken. To enable future FPOs to deal with the increasing volume of information on fire research and engineering their syllabus would need to be expanded if, for example, they are to continue to offer advice on all the fire safety implications of modern building technology. As currently trained, FPOs would be less able than BCOs to deal with problems of fire safety at the leading edge of fire engineering, especially in the context of advanced or innovative design and construction. However, the Fire Service College is moving towards revisions of curriculum and syllabus which would enable it to diversify its methods and to fit into a broader network of education and training. It was at the creation of this broader network that the report aimed in recommending the publishing of a basic text and the creation of an educational structure common to all three groups of participants.

Against this background of education diversity, the move to extend the scope of regulations and the role of the professional judgement in enforcing them led directly to two recommendations on the education of the various participants. Recommendation 2 of the report was a plea for a basic text for professional development in the subject and Recommendation 9 advocated a network of combined courses for the various groups. So far these educational proposals have met with a very encouraging response in the form of at least three initiatives. In May 1991 the Home Office circulated the draft of the proposed national core curriculum prepared by the Fire Safety Studies Working Group, itself a very useful new body with representatives from the professional bodies of architects, surveyors, building control officers, fire engineers and scientists.

June 1991 saw the publication of Paul Stollard and John Abrahams's excellent book, *Fire from First Principles* [3], as a very worthy first step towards the basic text recommended in the report. This book presents a collection of the keynote papers and is another first in collaborative education. This is a very good start but it must be emphasized that it is **only** a start and so much depends on the success of this educational effort that no apology is made for returning to the matter several times in this introduction.

Success with the procedural recommendations has not been quite so gratifying to date, but nor has it been negligible. The Department of the Environment (DoE) produced the procedural guidance document as suggested in Recommendation 1 [4]. Recommendation 7 on the issue of a

Part B Compliance Certificate has been included as no. 15 of the 1991 Regulations. With the exception of prisons, Part B of the Regulations now applies to all building types and is extended to cover Fire Brigade access and facilities, in the spirit of Recommendation 10.

Following up on Recommendation 6, the Home Office is encouraging Fire Prevention Officers to observe the consultation procedures to ensure that, wherever practical, advice to applicants should be confirmed in writing, distinguishing between the requirements of legislation and their recommendations that the applicant is free to follow or disregard.

In other areas, government has not yet had the opportunity to implement the Report's recommendations. For example, the enforcement of the new Regulations is the question of rescinding the requirements in local Acts once their content has been covered in national Building Regulations. Take as an example the controlling of compartment size and periods of fire resistance in car parks, a matter often covered in local Acts. The standard at present enforced under local Act provisions by local authorities may, in some cases, be higher than those in the guidance in Approved Document B. Yet until the local Act provisions are rescinded, the Building Control Officer may choose to enforce the higher standard, although it is to be hoped that the national standards will prevail.

The level of provision in Approved Document B is intended as guidance to both the applicant and the BCO. The applicant may choose to comply with the Regulation in some other way or, alternatively, the BCO may consider that the standard in the Approved Document is not appropriate in the circumstances of the particular case under consideration. In both cases, the necessary discussions and negotiations, which will need to take place, will be based on either first principles or other authoritative documentation. Until Recommendation 2 is more fully implemented, with the publication of further comprehensive design guides covering all aspects of fire safety and most types of buildings, and until the educational development in Recommendation 9 of building designers, BCOs and FPOs has been encouraged by the launching of many more professional development courses on the Queen's University model, the number of disputes and degree of dissatisfaction with the system may well increase.

Recommendation 3 proposed an arrangement whereby the BCO and applicant as soon as they identify an issue on which they appear to be heading for an irreconcilable disagreement should be able – in advance of the Building Regulations application – to make an early formal approach to the DoE to have the issue determined. The need for such an arrangement will be increasingly felt with the wide scope and deeper scientific base of the fire safety regulations.

The detailed changes in the new Regulations include efforts to simplify and extend guidance on a variety of fire safety matters. The extent to which these succeed will be discovered eventually in practice, and that will be closely monitored by the DoE and BRAC in the next year or two. It must be hoped that they are largely successful, for the field has broadened and deepened such that everyone needs as much help as possible. As evidence of this broadening, the Building Regulations, Part B, now apply to all buildings and have changed in status from 'mandatory' to 'guidance'. This apparent relaxation is more a recognition that the art and science of designing for fire safety is no longer susceptible to simple-minded prescriptions and indeed the Approved Document states that more of the new guidance is based on fire safety engineering as an alternative to rules of thumb, although these respective sources are not always clearly defined or evident in the guidance.

One thing is clear, rules are giving way to guidance and fire research, fire science and engineering are augmenting the empirical approach to an increasing degree. Fire science is advancing rapidly, and fire engineering is acquiring professional recognition. These are background considerations to the problems of the architects, Building Control Officers and Fire Officers, who have to deal with fire safety on a day-to-day basis and to collaborate effectively with their necessarily very different backgrounds. First, they must know about each other; secondly, they require to have a common core of knowledge; and finally, they must respect their differences as well as their common aims. So, in the educational effort skill-gaining is not an aim, while information-giving *is* to a degree, but most important of all is attitude-forming.

The difficulties surrounding this subject should not be underestimated: there is a very considerable range of practical and theoretical knowledge, together with the natural and behavioural aspects of fire safety and the rapid advances both in the pure science and its application in an increasing number of large, novel buildings. All these components will challenge the system in its effort to be flexible yet structured, and comprehensive yet manageable. Above all, it must be clear to each class of participant that the central objective is the safety of the building user, by design, by building, by use and management and, finally, by rescue. In any objective discussion of the subject, it is all too easy to forget just how terrifying fire in a building can be; how swiftly flame and smoke spreads; and how it is often little things like whether a door is open or closed that can make the difference between life and death.

The problem of educating people to exercise better judgement in the face of expanding knowledge is not confined to new building. Although attempting to preserve the simple ideas that Building Regulations apply only to new building – and only at the time of construction and not during their use – some recent changes are making this neat distinction less clear. In the UK the removal of Crown Exemption from the National Health

Service estate, and the national condition survey aimed at increased statutory compliance in colleges and polytechnics under the new funding regime, are examples of actions which have brought fire safety of the existing stock to the forefront of minds previously blissfully unaware of the issue. Managers, bursars, trustees, and the like, are increasingly aware of their responsibilities not just to ensure that any proposed new building alteration or extension is properly designed, but that (by a simple further thought) all their existing buildings should be, at least, surveyed and that fire safety upgrading should be part of their future business plans and budgets.

Therefore the upgrading of existing, occupied buildings to improve fire safety is likely to increase – and it raises some special problems because another simple idea so important to retain is the view that, once the contractor takes possession of the site, the programming of all the building operations and the management of and insurance against all risks are his sole responsibility until he hands back the site on completion. But unlike most other building operations, the sequence and management of the work on, for example, an occupied hospital cannot be left to the building contractor. The contractor does not know the medical, nursing and estate management constraints which are needed to ensure the continuous operation and maintenance of a working hospital, for the benefit of patients and visitors; nor can he predict or dictate the rate of spending on what may become a series of separate but interdependent capital projects spread over five or six years. The contractor cannot unilaterally plan his works compound and delivery system within an already crowded site alongside life-and-death provisions such as accident and emergencies departments, ambulance and Fire Brigade access. And yet, if all such arrangements are not made beforehand and maintained throughout the works, their late provision may prove costly or even dangerous. Special difficulties arise also from the fact that any work in an occupied building makes conditions worse before they can get better; for example, fire danger increases during construction, even if the end-result reduces it, and tight hospital sites require traffic management to reconcile the builder's work space with normal hospital activities, while maintaining fire-fighting access and fire escape routes.

The strategy and techniques to meet and minimize these difficulties fall to the design team and client liaison personnel from the earliest stage of planning a major refurbishment through to its completion; and the implications of the strategy in terms of sequence or work, restrictions on access and special provisions for collaborating on security must be built into the contract at the outset if delay, dispute and cost penalties are to be avoided.

The question of responsibility and liability must change as compliance with statutory requirements becomes more a matter of judgement and less one of merely following simple rules. In an increasingly complicated and litigious world, while the building control authority and the fire authority are the statutory bodies responsible under their various Acts for ensuring compliance, there is no body of case law which relieves the designers, builders, managers and owners of buildings of their responsibilities should their work be held to have contributed to death or damage. Fortunately, the law does not penalize professionals for being wrong, but rather for negligence, otherwise the industry would be paralysed by uncertainties. Avoiding professional negligence implies two things: keeping abreast of the state of the art or, at least, sound common practice; and applying knowledge diligently in the judgements that are made. In a fast-developing field, such as fire safety engineering, tracking the current state of the art is going to be increasingly difficult and the circumstances surrounding the judgements that are made about, for example, trade-offs between active and passive safety measures in innovative designs may be equally difficult to define.

REFERENCES

1. Bickerdike, Allen (1990) Report, *Fire and Building Regulations*, HMSO, London.
2. Allen, W. (1987) *Fire and the Architect*, Fire Research Lecture, FRS, Borehamwood.
3. Stollard, P. and Abrahams, J. (1991) *Fire from First Principles*, E. and F.N. Spon, London.
4. Department of the Environment, Home Office and Welsh Office (1992) *Building Regulations and Fire Safety: Procedural Guidance*, HMSO, London.

1
Fire science

Dougal Drysdale

1.1 INTRODUCTION

Fire science can be regarded as that subset of the sciences that has a direct contribution to make towards fire safety and fire protection. Its development over the past two decades has followed the growing awareness of the need for a more rational approach to all aspects of fire safety, and it provides the basis on which the practice of fire safety engineering can be built. Its principal objectives relate to life safety and property protection. Although these two areas are frequently considered separately, measures taken for one will normally be beneficial to the other. Nevertheless, almost all regulations and standards relate to life safety.

1.2 THE CRITICAL STAGES OF FIRE

To understand the role that fire science has to play, the critical stages of a fire, and how they relate to the objectives of life safety and property protection, will be considered. Although no two fires are exactly the same, key features which are common to all can be identified. Assuming that a fire is not extinguished, it will exhibit three main stages [1]:

1. the growth period;
2. the fully developed stage; and
3. the decay period.

These are identified in Figure 1.1. The duration of each stage will depend on a number of different factors. The growth period commences with ignition and ends with 'flashover', which is best thought of as the transition between the first and second stages. The duration of the fully developed stage will depend on the amount of fuel available and on the rate of burning, which is determined largely by the rate at which air can enter the fire compartment through openings such as doors and windows. As the fuel is used up, the fire will begin to decay.

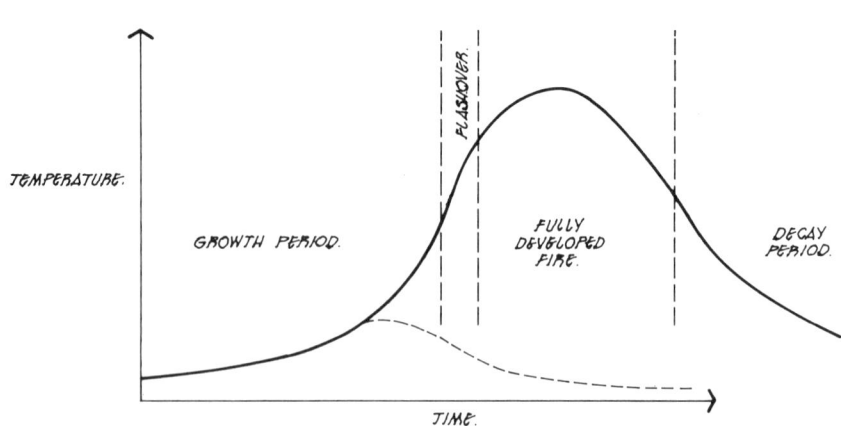

Figure 1.1 Course of a compartment fire, expressed as an average gas temperature as a function of time (broken line represents depletion of fuel before flashover has been achieved).

Strategies for fire safety can be based on the prevention of ignition, and/or the prevention or delay of flashover. Once flashover has occurred in the room or compartment of origin, the fire has the potential to spread to the rest of the building and cause major damage, both to the contents and the structure of the building. Furthermore, all those still in the building are directly threatened.

1.3 PREVENTION OF IGNITION

The methods of ignition prevention can be summarized under the following four headings:

1. Control/eliminate ignition sources.
2. Control/eliminate materials that are easily ignited.
3. Use non-combustible materials.
4. Use materials of 'low ignitability'.

While each of these may be obvious, common-sense solutions, the quantification necessary for rational decision making requires a fundamental understanding of the ignition process, including the interaction between the ignition source and the combustible material.

Combustion involves a complex series of rapid chemical reactions between fuel and oxygen (usually from the air), releasing heat and light. Flame is the visible manifestation of this process, indicating that the fuel is in the gaseous state. The 'gasification' of a combustible solid occurs if it is heated sufficiently for it to degrade chemically to produce flammable vapours. Its temperature (more specifically, its surface temperature) must be raised to the firepoint, which corresponds to the minimum condition at which ignition of the vapours (e.g. by a 'pilot' flame) will lead to sustained burning of the solid. The vapours will ignite spontaneously if the surface temperature is raised significantly higher than the firepoint. (Typically the firepoint corresponds to a surface temperature of 300–350° C, while the spontaneous ignition temperature will be at least 100° C higher.)

It is worth drawing attention to the fire triangle, which is often invoked to emphasize that fire can occur if fuel, air and oxygen (the sides of the triangle) are brought together. The removal of any one of these three, namely fuel, oxygen or heat, will lead to extinction, the most appropriate in the present context is the removal of heat (cooling) by the application of water. Although this is a useful concept, it can be seen from the above discussion that it is an oversimplification; for example, it does not reveal that the mechanism of heat transfer (conduction, convection and radiation) is the dominant factor not only in ignition, but in all the other stages of fire as well.

The ignition process thus involves the transfer of heat from an ignition source to a combustible material. How quickly the latter will reach its firepoint and begin to burn will depend on its thickness and its thermal properties – its thermal conductivity (k), heat capacity (c) and density (ρ). Materials which have low values of the product $k\rho c$ (known as the thermal inertia) respond quickly to an imposed heat flux and ignite rapidly if an ignition source is present. Thus standard polyurethane foam (for which $k\rho c = 1000\ W^2.s/m^4.K^2$) will ignite and continue to burn if a small flame is applied for only 2–3 s, while a typical solid plastic more than 5 mm thick (such as Perspex) has $k\rho c = 320\,000\ W^2.s/m^4.K^2$, and may take 40–50 s to ignite under similar circumstances.

In general, combustible materials of low density (e.g. insulating materials, etc.) can be ignited very easily. Similarly, thin sections of high-density materials (e.g. wood shavings, cotton fabrics, etc.) can be heated to their firepoints very rapidly. Ignition may be inhibited by the application of suitable fire-retardant treatments, but this will not render the materials non-combustible. They will still ignite if the ignition source is large enough, and they will certainly ignite and burn when involved in a compartment fire in which there will be a high level of background radiative and convective heating.

Once sustained burning has commenced, the heat transfer from the flames to the fuel will gradually increase the surface temperature until the maximum rate of burning has been achieved.

1.4 PREVENTION OR DELAY OF FLASHOVER

The prevention or delay of flashover is a much more complex issue than the prevention of ignition. To understand how either may be achieved requires some understanding of almost all of the various 'fire processes', including ignition, flame spread, rate of heat release, the buoyant movement of flames and, above all, the associated heat transfer mechanisms. Once an item of combustible material has been ignited, flame will spread over the available surface, increasing the area of the fire and consequently increasing the overall rate of burning. The latter is determined initially by how much heat is transferred to the surface of the fuel from the flame burning above the surface.

If the item is in an enclosed space, then the presence of the ceiling has a profound effect on the course of the fire. The ceiling becomes heated and hot combustion products form a layer in the upper part of the enclosure, the net effect being that a significant proportion of the heat of combustion (which, in the open configuration, is lost to the atmosphere in the buoyant plume) is radiated back to the fuel surface; this has two effects:

1. The rate of spread of flame over the item first ignited is increased.
2. A much higher rate of burning is achieved.

This is illustrated in Figure 1.2, which shows the results of a laboratory experiment in which slabs of polymethylmethacrylate (Perspex or Plexiglass) were allowed to burn in an open and in a confined situation. In the latter, 'flashover' may be regarded as the period of rapid acceleration of the rate of burning (expressed in terms of the rate of mass loss per unit surface area in grams per square metre per second); this should be compared with Figure 1.1.

In general, flashover is likely to occur once the size of the fire – measured in kilowatts or megawatts – exceeds a critical value relevant to the geometry and ventilation of the compartment of origin. Radiation feedback from the upper part of the compartment increases as the size of the fire increases. When the upper layer temperature reaches 600°C (as it will shortly after the flames grow tall enough to reach the ceiling), flashover can be regarded as imminent. This is associated with heat fluxes of >20 kW/m^2 at floor level, but this will quickly be exceeded as the upper layer begins to burn. When this happens, the fire enters the 'fully developed' stage, all combustible materials will become involved and flames will emerge through the upper parts of any openings (doors, windows, etc.). The rate of burning will then be controlled, to a large extent, by the rate at which air can enter the compartment: this is the ventilation controlled fire.

Under circumstances where there are large window openings and the fuel is limited in quantity, in particular, it has a low surface area exposed to the fire, the rate of burning can remain relatively low and is controlled by the area of the exposed fuel surface. This is the **fuel controlled fire**, the temperature being moderated by the excess air drawn into the compartment during the burning process.

Prevention or delay of flashover

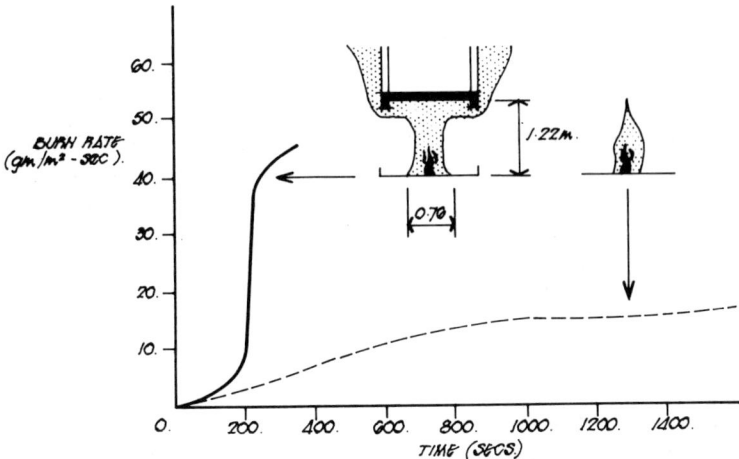

Figure 1.2 Effect of enclosure (solid line) on the rate of burning of a slab of polymethylmethacrylate (9.76 × 0.76 m); the unenclosure burning is shown as a broken line. (Source: R. Friedman, Behaviour of fires in compartments, *International Symposium on Fire Safety of Combustible Materials*, Edinburgh University Press, 1975, pp. 100–13.)

If there is inadequate ventilation available to the fire during the growth period, then the fire may starve itself of oxygen and fail to reach flashover. It may self-extinguish or it may continue at a low rate of burning or as a smouldering fire. Such conditions can be extremely hazardous as over a period of time the enclosure can fill with a flammable smoke. This can erupt into a vigorous fire if a new supply of air is created by the breakage of a window or the opening of a door. The latter case is referred to as a backdraught and is a particular hazard for fire fighters when they carry out routine searches for trapped persons.

A number of factors contribute to rapid growth to flashover; these include:

1. Tall items of combustible material.
2. Large expanses of combustible partitions and wall lining materials.
3. Items of large surface area which can be ignited easily and spread flame rapidly.
4. Items which can achieve a high rate of burning in a short period of time.
5. Building geometry (e.g. small compartments, low ceilings).

The significance of these factors is clear when interpreted in terms of the underlying science, some of which has been outlined above.

Some of the most important advances that have been made in fire science over the past ten years have led to a better understanding of the flashover process. Flashover represents the start of a quasi-stable phase of the fire when flaming is no longer localized, but occurs throughout the enclosure as the fuel vapours mix with the incoming air in a highly turbulent fashion. The amount of ventilation controls the rate of burning while the duration of the fire is determined by the quantity of fuel present. The post-flashover stage of the fire is of significance to the architect as it is during this period that maximum temperatures will be attained and the fire resistance of elements of structure will have to be taken into account. (In the final cooling period the fire decays as the available fuel is consumed.)

1.5 LIFE SAFETY

Various studies of fire fatalities in the UK and the USA have shown that the majority (>70%) can be attributed to the inhalation of smoke, the principal toxic constituent of which is carbon monoxide (CO). Very few victims die as a result of direct fire exposure, and although the bodies of approximately two-thirds of all fatalities are found to be very badly burned, these injuries in the majority of cases have been sustained after death. The presence of carboxyhaemoglobin in the blood of these victims, and the absence of burning of the upper respiratory tract, are indications that death occurred before direct exposure to the fire.

Smoke is best defined as the gaseous products of combustion in which small solid and liquid particles are dispersed; it contains burnt, and partially burnt, products formed in the flame, as well as some products that are given off by the chemical degradation of the fuel. The composition is complex, but except in a few unusual situations the toxicity can be attributed to the presence of carbon monoxide. However, other constituents such as acrolein (typically from wood) and hydrogen chloride (e.g. from polyvinyl chloride, PVC) can irritate the eyes and upper respiratory tract at very low concentrations and seriously impair a person's ability to reach a place of safety. All smoke should be considered as dangerous.

Smoke is formed as the hot products from the flame rise as a buoyant plume, entraining air as it moves upwards and thus increasing in volume. Indeed, air is by far the largest constituent of smoke, and therefore in any attempt to estimate the rate of smoke production it is necessary to assess the rate of air entrainment. This depends on the size and intensity of the fire, in particular, its perimeter and rate of heat output, and the height of the rising column before it reaches the ceiling, or enters the smoke layer under the ceiling [2]. In fact, due to the large number of variables, it is almost impossible to calculate accurately the rate of smoke production. The architect must appreciate that the larger the fire (in particular, the larger its perimeter), the greater will be the rate of smoke production.

Sprinkler systems are traditionally designed to limit a fire to an area of $9\,m^2$ (12 m perimeter); in a spinklered building it is normally assumed that this represents the largest probable fire area when estimating the rate of smoke production.

The appearance of smoke reflects its constituents and it will vary from a very light colour, typical of free burning wood fires, to a deep sooty black, typical of smoke from fires in which the fuel is oil or a hydrocarbon polymer (e.g. polyethylene). The 'density' of smoke depends on the concentration of particles contained in it: the greater the density, the greater the potential of the smoke to reduce visibility in escape routes. However, the ability to escape depends not only on visibility, but also on the physical and psychological condition of the individual who is attempting to escape. Moreover, if the smoke is irritating to the eyes and/or throat, movement can be impaired and escape delayed sufficiently for the individual to be exposed to a lethal dose of the toxic species present (i.e. mainly CO). Consequently, it is essential to maintain smoke-free escape routes.

The architect must consider all smoke as hazardous – indeed potentially lethal – although the toxicity will vary depending on the nature of the fuel, the conditions of burning and the amount of dilution that has taken place. All carbon-based materials (in effect all combustible materials) yield carbon dioxide as a principal product, with carbon monoxide as a product of incomplete combustion. Other toxic gases (some of which are more toxic than CO) are formed, but their concentrations are generally low in comparison to CO which remains the most hazardous species present. Nevertheless, many other toxic species, such as hydrogen chloride (e.g. from PVC) and hydrogen cyanide (e.g. from wool and polyurethane), are produced: their combined effect is poorly understood, and no general statement can be made about the effects of these 'toxic cocktails'.

It is more important for architects to be aware of why certain materials, or types of material, are hazardous. Much is said about the fact that polyurethane (PU) foam produces hydrogen cyanide. However, it produces carbon monoxide in much larger quantities. The fire in the Woolworth's store in Manchester, in 1979, involved a large quantity of polyurethane foam and caused the deaths of 10 people. It is often quoted to illustrate 'the lethal nature of the smoke produced by burning polyurethane', the implication being that the smoke from PU foam is much more toxic than that from other materials. This may or may not be true as small-scale toxicity tests are not a particularly reliable method of assessing the toxicity of smoke produced in real fires.

What is certain is that it was the rate at which the fire spread, generating enormous quantities of dense, black smoke in a short period of time that led to the large life loss. Those who died were close to a place of safety, but they were unable to reach the escape stairs because of the very rapid spread of the smoke. Hydrogen cyanide would certainly have been present in the smoke, but it is almost academic whether it was a significant toxicant: it was the rate of fire development that was responsible for the deaths.

In any type of fire scenario, life safety is dependent on early warning and efficient evacuation. People must be out of the room of origin before the fire has grown to more than about 100 kW in size (depending on the size of the room or compartment), and the occupants must be out of the building before flashover in the room of origin has occurred. If t_u is the time to reach flashover in the room of origin, then for safety of the occupants of the building, the following can be written:

$$t_u > t_p + t_a + t_{rs}$$

where t_p is the time that the occupants take to perceive that a threat exists, t_a is the subsequent delay before they start to move to a place of safety, and t_{rs} is the time it takes them to reach a place of (relative) safety. The time to perceive the risk can be reduced by automatic detection and alarm systems. The time to take action can be reduced by good instructions, staff training, etc., and the time to reach a place of safety can be reduced by ensuring good, clear means of escape. The time available for escape (i.e. t_u), which was much too short in the Manchester Woolworth's fire, can be increased by delaying the onset of flashover, or better still, by preventing it occurring. As discussed above, this can be achieved through careful selection and control of the combustible contents and linings, and perhaps by careful design of the building geometry and layout.

1.6 PROPERTY PROTECTION

Smoke damage to a building and its contents can be severe. The structure can normally be cleaned successfully, but the corrosive nature of smoke can cause irreparable damage to electronic circuits and control equipment. In the telecommunications, microelectronics and computer industries a relatively small fire (in terms of area directly affected) can result in a very high loss due to the high sensitivity of the equipment and products to corrosion.

However, if a fire is not extinguished at an early stage and it passes beyond flashover, it may spread to involve a significant part of the building and over a period of time the extreme heat can cause major structural collapse. Temperatures in excess of 1000° C are not uncommon in a fully developed fire and, given sufficient time, the elements of the structure will reach temperatures at which they will suffer significant loss of strength. As a rule of thumb, steel may be said to have lost about two-thirds of its strength by the time it has reached 600°C. Unless protected by fire-resisting cladding, this may be reached after 15–20 min exposure to a fully developed fire, depending on the mass of the section. Reinforced concrete may be regarded as a more resistant material, but as its tensile strength depends on the strength of the steel reinforcing bars, these must be protected by a sufficient thickness of concrete to provide enough thermal insulation to the steel. Surprisingly, although timber is combustible, it is a very good structural material because it burns relatively slowly: provided that structural timbers are oversized so that a 'sacrificial' outer layer can burn away without the load-bearing capacity of the element falling below its design requirement, they can provide a considerable measure of fire resistance. (Note that bricks provide one of the best fire-resistant materials as they have already been fired at high temperatures during manufacture; however, the stability of brick walls and supporting structures will depend on how carefully they have been designed and constructed.)

The rate at which heat is released, combined with the duration of the fire, may be considered as the main elements in the definition of 'fire severity'. An understanding of the factors which determine the level of heat production enable an estimate to be made of the potential of the fire to cause damage to a structure. In a compartment fire the rate of burning is controlled by the rate at which air can enter the fire compartment, while its duration depends upon the quantity of fuel present.

The quantity of the potential fuel within a building is described as the fuel (or fire) load of the building, and will include the combustible components of both the fabric of the building and its contents. An estimate of the fuel load can give rough guidance on the fire resistance requirement. However, it is very difficult to establish accurately due to the multiplicity of materials which are likely to be involved. The architect should be aware of the fact that the fuel load will depend on the building occupancy; thus the fuel load of a large distributive warehouse will be much higher than the fuel load of a sports centre of similar size. Data are available on the fuel load distributions to be found in different types of occupancy; in general, the fire resistance requirements are dictated by the occupancy type.

However, as discussed above, it is not just the amount of the fuel which will influence the rate of heat output, but its arrangement can play a dominating role. In essence, the greater the surface area of fuel exposed, the greater is the potential for rapid fire development. Its location is also important as a fire close to the walls or in a corner will develop more rapidly since there will be less entrainment of air into the buoyant gases and the temperatures under the ceiling will rise more rapidly. Under these circumstances, if the wall and ceiling linings are combustible, they will contribute to the developing fire at an early stage.

1.7 APPLICATION OF FIRE SCIENCE TO FIRE SAFETY

Fire science has given us a greater insight into the mechanisms of ignition, fire growth and fire spread, as well as into the problems associated with smoke movement and toxicity. It is also of value when it comes to examining the implementation of measures for improving the fire safety of the occupants of a building. Indeed it must be applied if significant advances are to be made in the engineering of fire safety. Two main areas of application can be identified:

1. Selection of materials and the control of ignition sources.
2. The application of fire protection technology, which includes automatic detection, automatic suppression, smoke control systems, compartmentation and, to a lesser extent, structural fire protection.

Only detection and suppression will be discussed here.

Materials which are to be used in buildings must be selected carefully for the function they are to perform. However, unlike properties such as Young's modulus and thermal conductivity, the fire performance of a material cannot easily be assessed as it depends not only on the nature of the material, but also on the nature and severity of the fire. This fact has only recently been accepted by the regulatory bodies who, traditionally, require the result of single fire tests – representing single model fire scenarios – as a means of ranking materials for selection purposes [3]; these tests have covered:

1. combustibility;
2. ease of ignition;
3. rate of surface spread of flame;
4. rate of heat release;
5. the production of smoke and toxic gas.

In general, existing fire tests may be said to be discredited in the light of a new understanding of fire science and fire behaviour. The tests BS 476: Parts 6 and 7 continue to be used in the UK because experience tells us that they provide an 'acceptable' level of safety although they were developed for traditional materials (i.e. wood and wood products) and are unsuitable for synthetic materials, many of which melt and flow on heating, and may not be severe enough to identify those which will present a hazard when exposed to fires involving synthetic materials. However, the greatest problem lies in the need to have fire test methods which will be acceptable on a European-wide basis. At present each country within the European Community (EC) has its own test, or test protocol, for assessing the fire safety of combustible lining materials [3] but the results of these tests do not give consistent results [4]. Different ranking orders are obtained, and there is currently a major debate in progress on how this issue can be resolved, so that a material which is deemed to be acceptable in one European country can be marketed throughout the EC.

One solution which has gained favour is the development of a new generation of tests, which will give us data on materials and allow rational decisions to be made about their suitability for use in specific scenarios. This also requires us to be able to define the fire exposure to which the material may be subjected. At present, the cone calorimeter is regarded as the most promising apparatus for providing such data [5]; this measures, as a function of imposed heat flux:

1. the time to ignition;
2. the rate of burning (rate of mass loss);
3. the rate of heat release (measured by oxygen consumption calorimetry);
4. the time to maximum rate of heat release;
5. the carbon monoxide, carbon dioxide and smoke production.

There are already a number of simple models for predicting fire spread on wall lining materials, using data from the cone calorimeter. Much work remains to be done before this technique is acceptable throughout the EC, but inevitably this approach will form the basis for the future method for materials selection and will be of benefit to both designers and regulators.

Traditionally the installation of automatic detection and suppression systems are prescribed in British Standards documents. These are based on experience of the use of such systems ('rule of thumb'), and are highly restrictive and often found wanting. Detection systems are normally based on the identification of a characteristic fire signal such as smoke, heat, carbon monoxide, etc. Here there is much scope for improvement, using the new knowledge of the behaviour of fire plumes and the associated flows of hot combustion products under the ceiling. The choice of detection or suppression system must be based on the size of fire that has to be detected (or the maximum size of fire that can be tolerated). If it is known how quickly a fire can grow to this critical size (measured in kilowatts), then it is possible to specify the characteristics of the detector necessary to detect the fire early enough for an alarm system to be operated or a suppression system activated.

Currently there is a National Fire Protection Association (NFPA) standard which describes a method for specifying the response time of a heat detector and the spacing of the heads necessary to give adequate warning for a variety of fire types [6]. This is a first attempt to translate the considerable body of research knowledge in this field into a format that has direct application to the fire safety engineer. It is simply a matter of time before comparable advances are made in other areas [7].

REFERENCES

1. Drysdale, D. (1985) *Introduction to Fire Dynamics*, Wiley, Chichester.
2. Thomas, P., Hinkley, P.L., Theobald, C.R. and Simms, D.L. (1963) *Investigations into the Flow of Hot Gases in Roof Venting*, Fire Research Technical Paper No. 7, HMSO, London.
3. Troitzsch, J. (1990) *International Plastics Flammability Handbook: Principles, Regulations, Testing and Approval*, 2nd edn (trans. J. Haim), Hanser, Munich.
4. Emmons, H.W. (1974) Fire and fire protection. *Scientific American*, no. 231, 21 July.
5. Babrauskas, V. and Grayson, S.J. (eds) (1992) *Heat Release in Fires*, Elsevier, Amsterdam.
6. National Fire Protection Association (1984) *NFPA 72E Standards on Automatic Fire Detectors*, NFPA, Boston, Mass.
7. di Nenno, P.J. (ed.) (1988) *The SFPE Handbook of Fire Protection Engineering*, NFPA/Society of Fire Protection Engineers, Boston, Mass.

2
Fire safety engineering

Paul Stollard

2.1 INTRODUCTION

The design process can be viewed as the attempt by an architect to satisfy a series of objectives; it is the search for a physical solution to a given set of problems. These objectives will include aesthetic, functional, technological and economic issues. If the building is to be successful, then these objectives will have to be integrated into a coherent and balanced whole. Among the technological objectives will be those of fire safety.

Fire safety is normally considered to cover both the safety of people and of property, both in the building concerned and in the surrounding area. Therefore the fire safety objectives for the architect will be twofold: life safety and property protection. Other objectives might sometimes be relevant, but these will normally be only variations on or combinations of these two principles. For example, in the fire safety design of hospitals the maintenance of the service is considered an objective (to avoid consequential life loss due to postponed operations and treatment), yet this is only a variation on life safety and property protection rather than a completely new objective.

In designing to ensure life safety the architect is seeking to reduce to within acceptable limits the potential for injury or death to the occupants of the building and for others who may become involved. The objective of property protection is the reduction to acceptable limits of the potential for damage to the building fabric and contents. The architect will be seeking to ensure that as much as possible of the building can continue to function after a fire and that the building can still be repaired. The building should also remain safe for fire-fighting operations during the fire. The risk to adjoining properties will have to be considered, as well as the wider risk of possible environmental pollution.

The two principal products of combustion relate to these two objectives and, in very crude terms, life safety can be seen as protecting people from smoke, while property protection concerns keeping heat away from the building. This gross oversimplification provides a succinct summary of the objectives which architects must fulfil and the dangers they must avoid.

2.2 TACTICS

With the fire safety design objectives of the architects agreed as being life safety and property protection, it is possible to begin delineating the specific tactics which the designer can use to achieve these objectives. Such tactics have to contribute to saving lives and the protection of property; failure will lead to death and destruction. Five tactics can be clearly defined, and these are:

1. Prevention – ensuring that fires do not start, by controlling ignition and fuel sources.
2. Communications – ensuring that, if ignition occurs, the occupants are informed and any active fire systems are triggered.
3. Escape – ensuring that the occupants of the building and the surrounding areas are able to move to places of safety before they are threatened by the heat and smoke.
4. Containment – ensuring that the fire is contained to the smallest possible area, limiting the amount of property likely to be damaged and the threat to life safety.
5. Extinguishment – ensuring that the fire can be extinguished quickly and with minimum consequential damage to the building.

If the above tactics are considered in order, then the first is obviously prevention – only if this fails need the other tactics be attempted; if fire prevention is successful, the others need not be attempted. However, as fire avoidance will inevitably fail at some stage during the life of the building, provision *must* be made for the other tactics.

Communication by itself, even if totally successful, cannot save lives or protect property, but its key role in ensuring fire safety means that it must be considered as one of the five tactics. If communication is successful, then escape and extinguishment can be attempted; but if it is unsuccessful, then only containment remains as an available tactic. Containment by itself will not achieve the twin objectives, the success of containment can only win more time for communication to be successful.

These five tactics provide the fundamental framework within which the architect should be working. They can be considered as a flow chart leading to the success or failure of the objectives (Figure 2.1). A building designed with adequate consideration given to these five tactics will offer an acceptable level of fire safety. Each of the tactics will be the subject of one or more chapters in this book. The relationship between these fundamental fire safety tactics and the current legislation with which

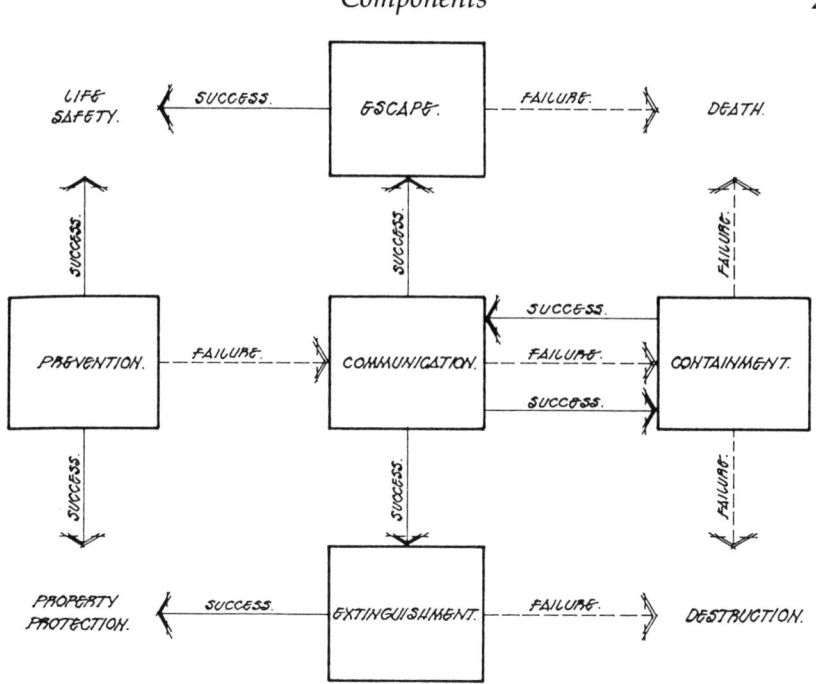

Figure 2.1 Matrix of tactics and objectives.

architects have to comply is an interesting one, and this will be considered in section 2.4 on acceptability and equivalency.

2.3 COMPONENTS

Fire safety can be viewed as a hierarchy (Figure 2.2) with the objectives achieved through the successful implementation of the tactics – and these tactics relying on the right mixture of practical safety measures or fire safety components. These components are the weapons that the designer can use tactically to achieve fire safety: they are the building itself, its furniture, fittings and occupants. The number of components is limitless and depends solely on how they are categorized; they are what are actually built or installed – the fire doors, sprinklers and escape stairs. It is essential not to confuse such specific components with the more general tactics and objectives which the architect must follow. Compartmentation is a valuable weapon to the designer, but if used without understanding, it does not constitute an effective tactic for fire containment or achieve the objectives of property protection. Fire safety design is the integrated use of components to achieve the designated objectives. In order to achieve this, it is essential that the designer has an understanding of the principles underlying fire safety.

Fire safety engineering

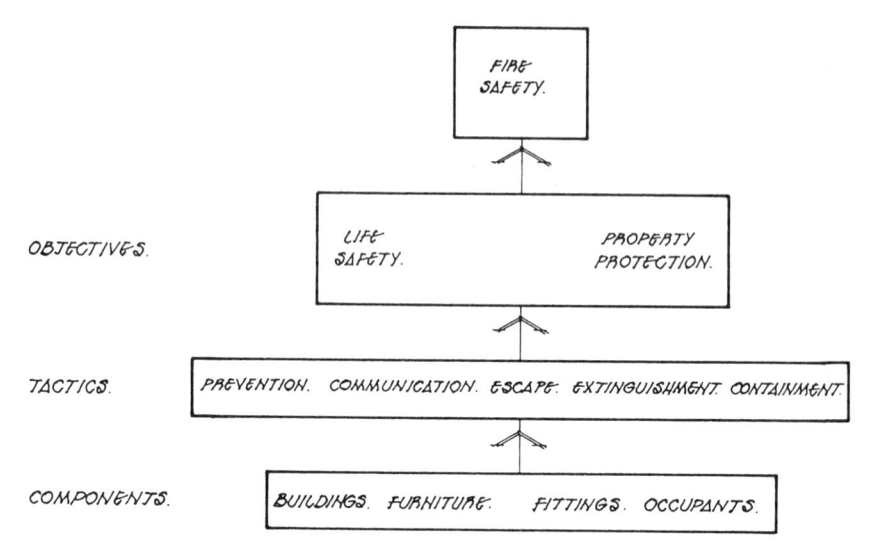

Figure 2.2 Objectives, tactics and components hierarchy.

With the necessity to assess the fire safety of all workplaces under the European Framework (89/391/EEC) and Workplace (89/654/EEC) Directives, much thought has been given to the components which need to be included in such assessments. One particular example is the assessment scheme developed for hospitals which considered both fire precautions and fire risks. Fire safety precautions are assessed as a list of components grouped under the headings of the five tactics already mentioned, as follows:

1. Prevention
    Management
    Training
    Housekeeping
    Signs and fire notices.
2. Communications
    Alarm and detection systems
    Observation.
3. Escape
    Single-direction escape (Stage 1)
    Travel distance (Stages 1 + 2)
    Refuge
    Stairways (Stages 3 + 4)
    Height above ground level
    Escape lighting
    Staff

    Escape bed lifts.
4. Containment
    Elements of structure
    Compartmentation
    External envelope protection
    Smoke control.
5. Extinguishment
    Manual fire-fighting equipment
    Fire Brigade
    Auto-suppression.

The assessment of these precautions is then measured against the list of risks identified. These risks are grouped under four headings each of which also has a series of components as follows:

1. Ignition risk
    Smoking
    Security
    Equipment
    Lightning
    Fires started by patients
    External sources
    Fire hazard rooms.
2. Fuel risk
    Surface finishes
    Textiles and furniture.
3. Patients – life risk.
4. Property loss and consequential harm.

In a similar fire risk and safety assessment scheme which has been developed to assist housing managers in determining priorities for improvement, a similar list of components was developed. However, it differs slightly to reflect the different building use. Again, the fire precautions are organized under the five tactics as follows:

1. Prevention
    Maintenance and fire safety
    Information for tenants
    Fire safety training for housing staff
    Gas and electricity
    Electrical sockets
    Storage.
2. Communications
    Smoke alarms
    Public telephones.

3. Means of escape
   Escape route
   Windows
   Fire resistance of shared escape route.
4. Containment
   Fire protection betweeen homes
   Lofts, ducts and cavities
   Inside walls and ceilings
   Outside walls.
5. Extinguishment
   Fire Brigade access.

In the housing assessment the risk factors considered are:

1. Ignition risk
   Heating
   Number of homes sharing escape route.
2. Fuel risk.
3. Life risk
   Tenants
   Height above ground.
4. Property risk.

With building types other than hospitals and housing, different components may need to be considered. These will include not only the obvious (such as fire extinguishers), but everything from the wall coverings to the management of the occupants which may be relevant.

Each of the components may contribute to one or all of the five tactics, and it is this complexity of interaction which necessitates a logical approach to the tactics of fire safety. There will also be interactions between the objectives, between the tactics, and between any of the individual components. For this reason, measures taken to reduce the fire risk or hazard cannot be viewed in isolation and the overall impact of any measures must be considered.

For instance, the provision of sprinklers in a building to improve the property protection may reduce the risk of a fire growing beyond certain limits. This restriction in fire size and rate of subsequent fire growth should reduce risk of structural failure and limit the amount of smoke produced. It should also increase the amount of time available for escape by containment of the fire. However, it will probably also reduce the smoke temperature and this will increase the possibility of local smoke logging. It may also reduce compartment pressures and hence increase the risk of smokelogging of staircases. These problems of smoke control may result in an increased risk of life loss. There is also the risk that the sprinklers may not function efficiently, and this may alter the risks to life and property.

The fire safety decisions are therefore complex ones and the designer has to be aware that changing one component or altering the emphasis placed on one of the fire tactics can have an effect on the probability of success in each of the two objectives.

In the more complex of designs, it may be necessary to attempt to consider such interactions quantitatively; but for most projects it is sufficient for the designer to be aware of the possible implications of his decisions.

2.4 ACCEPTABILITY AND EQUIVALENCY

Absolute safety from fire, where there is no risk whatsoever, is an ideal which it is impossible ever to achieve. The architect is never asked to provide such absolute safety, only to reduce the risks to property and people to a level which society regards as acceptable.

This acceptable level of safety has traditionally been defined through legislation. However, legislation tends to be produced as a response to particular problems or fires and it does not always offer a balanced or reasoned structure for fire safety. There is an argument that the whole history of fire safety legislation is simply a catalogue of responses to serious fires. It can be shown that not only is legislation enacted in response to disaster, but also that changes in building forms and technologies normally occur in response to disaster. In the case of the Bradford City football ground (1985) fire, the Home Secretary had announced within two days that higher safety standards were to be required at third- and fourth-division football grounds. Yet this was the first fire at a football ground in which a member of the public had been killed. Public reaction demanded such moves, and within two days of the fire six other football clubs had either closed stands or started to remove perimeter fences and provide exit gates. However, these reactions were not based on any rigorous assessment of risk, nor did the legislation form part of any general or comprehensive strategy for a common level of safety in all buildings. The vast majority of people who die in fires do so in their own homes, on average two a day in the UK, and yet because these are small incidents which rarely gain media coverage, the standards of fire safety required in domestic dwellings are possibly lower than in most other buildings.

With much of the legislation related to fire safety being introduced responsively following particular tragedies, the existence of a coherent fire safety policy can be queried. Society is happy to accept as safe all buildings in which the dangers have not recently been exposed by a serious fire. Compliance with the regulations which are in force at a particular time is assumed to provide an acceptable level of safety – even though that level cannot be objectively quantified.

There are alternative ways of measuring safety than showing compliance with legislative standards. It is possible to design against specified risk criteria; for example, designing to ensure that the probability is of a single death occurring once every thousand years and a multiple death once every million years. It is also possible to design against specified probabilistic or deterministic criteria – i.e. to ensure that people within a building are able to reach a place of safety in a time less than that for conditions within the building to become untenable.

It is obvious that there is a point beyond which increases in fire protection measures add to the cost in undue proportion to the added safety they provide. The fire safety designer must therefore achieve a balance between safety, economics and convenience. Acceptability must be discussed in view of the fact that absolute safety cannot be achieved and the 'law of diminishing returns' (i.e. a type of cost–benefit analysis).

The other important concept linked closely to acceptability is equivalency. Once architects are able to achieve an agreed acceptable level of safety by whatever combination of tactics they choose, then the importance of ensuring equivalence is critical. Equivalence between two different fire safety designs means that they achieve the same level of safety by different methods. This is sometimes described as a 'trade-off' – the concept that one safety measure is being traded-off for another. For example, does concentration on fire escape enable the architect to pay less attention to fire extinguishment, or do measures installed to decrease the possibility of ignition balance a decrease in fire containment measures?

Attempts to assess equivalency in terms of a single numerical value are difficult and can hide a number of contradictions. For example, one might regard the presence of sprinklers as providing an additional level of safety which would permit an increased escape distance. This would then perhaps reduce the annual loss of life with an occasional larger loss, resulting from the one in 50 times that the sprinklers fail. A strategy for equivalency must recognize the distinction between average and societal risk. Calculations of equivalence are therefore neither simple nor easy to quantify, and for this reason, they have not normally been explicitly incorporated into the legislative framework.

Many of the current Building Regulations are couched in terms of what is 'adequate' or 'reasonable'; however, these terms are not defined, except by reference to approved documents which prescribe how each component must be designed. There is no attempt to define either acceptable safety or a system of determining equivalency, yet by saying that the approved documents constitute what is 'adequate' and 'reasonable', then any alternative fire safety strategy offering equivalent or better levels of safety should be acceptable. If designers are to be able to satisfy the objectives of fire safety without compromising other objectives

2.5 TRADITIONAL AND HOLISTIC APPROACHES TO FIRE SAFETY DESIGN

(economics, aesthetics and functionality), they need to be aware of the full range of different, but equivalent, fire safety strategies.

The traditional approach to fire safety as espoused through the Approved Documents of the Building Regulations, has been to identify certain components and then to prescribe certain standards for these components. Such components in the current Building Regulations for England and Wales (1992) which apply to new buildings include:

- travel distances and routes
- loadbearing elements of structure
- roof construction
- separating walls
- compartment walls and compartment floors ← *mention also space separation*
- protected shafts
- concealed spaces and fire stopping
- internal surfaces
- stairways
- Fire Brigade access and facilities.

The Fire Precautions Act 1971, which covers existing premises as well as new developments in certain building types, also covers such issues as:

- staff training
- manual fire-fighting equipment
- detection and alarm systems
- emergency signs and lighting.

Such an incremental approach has already been criticized for limiting the design choice of architects, giving no guidance on aceptability and not helping in any calculation of equivalency. However, the most serious criticism of such an approach is the total neglect of some aspects of fire safety. Fire prevention is hardly mentioned and smoke control is not given the attention that it warrants. The traditional approach was to regard all these components of fire safety as somehow independent and to demand a 'reasonable' standard of provision in each of them. This limits the design flexibility of the architect and can lead to a resentment of the legislation. The architect starts to seek loopholes or ways to get round the legislation. Such an approach guarantees that the designers and the legislative authorities will be in opposition to one another.

The traditional approach also creates an artificial distinction between the requirements of the legislation which are normally supposed to concentrate on life safety, and those of the building's insurers which will be more concerned with property protection. Yet most fire safety measures will contribute in some degree to both life safety and property protection. The artificial separation of the two can lead to examples of both areas of overlap and gaps. Conflict can even be generated by the different priorities of the legislation and the insurers: in one Oxford shopping centre the insurers offered a reduced premium if the mall was sprinklered, but this was not done because of the increased risk to life that the architects felt would result.

The alternative approach to fire safety can be described as the holistic (or fire engineering) approach and is the one underlying the chapters in this book. In this, the building is considered as a complex system, with the fire safety design just one of the many interrelated sub-systems. The architect is faced with designing not just to satisfy a series of prescriptive standards, but designing also to achieve an acceptable level of safety. This will require an assessment of the equivalence of alternative fire safety strategies and the development of an integrated approach to fire safety. In the holistic approach issues like fire avoidance and fire communication can be given their due weight and the designer can fully exploit all the techniques of improving fire safety. The holistic approach demands an understanding by the designer of the fundamentals of fire safety, but it offers the opportunity to attempt unconventional ways of achieving compliance with the legislation. All the designer has to do is prove that what is being offered represents a level of safety equivalent to what has been defined as the acceptable level.

The holistic approach will begin with a recognition of the objectives of fire safety, life safety and property protection. Then the fire safety design must be prepared on the basis of an assessment of risk and an analysis of the protection offered by fire safety measures. The assessment of the risk will include the building risk, the ignition risk and the fire loads. The building risk will be related to the geometry and design of the building, and the ignition risk is dependent on the probability of ignition within the building (how it is used and who uses it). The fire loads can be calculated from the fuel available to burn, either in the fabric of the building or its contents. A subset of the fire loads are the smoke loads which will depend on the ability of materials to produce smoke and toxic gas, a critical feature in establishing life safety. There must also be a corresponding assessment of the safety offered through the design including the potential for escape, containment and extinguishment.

Such assessments of risk and safety may be either qualitative or quantitative. Qualitative techniques, such as points schemes, rely on an expert-based assessment of risk. The only such scheme widely used in the UK at the moment, assessing hospital fire safety, has already been mentioned. Quantitative techniques are also available such as fire growth modelling, smoke models, structural models, probabilistic analysis, structural response modelling, environmental testing and extrapolation of results, deterministic calculations, stochastic evaluation, fault tree, event tree and critical path analysis. However, at present such quantitative models are too complex in form, and too limited in scope, to be of significant value to the ordinary architect. It is hoped that more accessible quantitative models will be developed which can be easily used by the designer to enable him to assess the fire safety of his proposals.

The rest of the chapters in this book are structured round the holistic approach to fire safety, and each of the five tractics available to the architect in preparing a fire safety design is covered in one or more chapters. Although it is sadly not yet possible to offer a simple quantitative system which would permit a full study of the equivalency of alternative proposals, an understanding of the fundamentals will enable the architect to make approximate judgements on equivalency and on the acceptability of the proposed fire safety measures.

The gradual improvement of the Building Regulations may lead hopefully to a stage when the safety demanded by legislation will be coherent and balanced for every building type, with designers being asked to achieve levels of safety defined in terms of risk to people and property rather than in the performance standards of compartment sizes, door widths or travel distances. Throughout this book, and in any approach to fire safety based on first principles, architects, fire engineers and the statutory authorities must consider safety in this manner.

3
Fire prevention – designing against arson attack

Lawrence Johnston

3.1 INTRODUCTION

The act of arson is a crime resulting in losses due to fire destroying property, whether it is buildings or contents, or both. The financial cost of these fire losses presents a major problem in the UK, and for society, individual communities and insurers. The Association of British Insurers now recognizes arson as the greatest cause of fire loss in the 1990s. The setting up of the Arson Prevention Bureau in February 1991 was an attempt by the Home Office and the Association of British Insurers to play a key part in initiating and co-ordinating efforts to combat the growing menace of arson [1].

The Report of the Working Group (Home Office, Standing Conference) on the Prevention of Arson was published in December 1988 [2]. It highlighted the role of the architect in the chain of measures likely to reduce the risk of arson and urged architects to include preventive measures within their plans. The Report also recognized that 'in respect of arson prevention, it is persuasion and guidance, not changes in legislation, that offer the best way forward'. What can the designer do when faced with the problem of arson in new, purpose-built property?

To undertake the specific task of designing against arson attack in new buildings will necessitate an overview of several key factors which should be capable of evaluation at the outset. Whether the architect is in the traditional role of leader of the design team for the project or a single contributor to a multi-disciplinary management team, the consideration of design against arson attack should be a subject included at the start, not a matter to be added on at the last minute. If the architect can at the outset bring together the various interested parties, statutory bodies, insurance representatives and others, it could offer the client lower costs of fire insurance premiums, safer buildings and a reduction of abortive work and duplication. Too often it seems that the necessary steps to reduce the risk of arson attack are implemented at the end of the project. If increased costs result, the client can often become frustrated by having to pay for these extras, and critical of the architect for not having known about the need to undertake straightforward requirements to obtain minimum insurance cover. On occasion, some measures are taken to reduce fire risk, but because specific installations, such as sprinkler systems, are not up to an acceptable insurance standard, they have little or no discount effect on the fire rating of the property. The client is quite naturally perplexed that initial efforts are not offering financial rewards.

The context of the site location for the project can be a significant indicator in the arson risk equation. Recorded losses due to arson attack point to certain patterns of those areas where there is a strong likelihood of incidence. These include: areas of urban deprivation; areas where there is high unemployment in the community; and isolated or exposed sites, sometimes on the periphery of declining residential or industrial estates. Locations which may lie along segregated residential urban estates tend to be the areas where community tensions are strongest and street violence can erupt; the adjoining property is at risk of arson attack and setting fire to the surroundings is an attraction. Where the end-use of the new building is known and falls into a particular category of building type, there are valuable pointers from past records which can assist the design teams to assess the risk of arson attack. One example would be the construction of a new school in an existing deprived residential area. The very existence of the new building in this context may make it a likely target for arson attack: 'The incidence of arson in schools is exceptionally high. The average risk of a deliberate fire in a school is approximately one in thirty-seven per year. The overall risk in secondary schools is about six times as great as in primary schools and fire damage in secondary schools is more expensive to repair' [2]. The design team and the client can take into consideration whether the arson attack could arise from present or immediate past pupils acting out a grudge, or from other children not part of the new school who view the building as an invasive threat. They must make use of all published guidance from official sources [3].

3.2 THREE LINES OF DEFENCE

When considering more general factors for new build, the relationship between the property and its immediate external space is significant. If the building is to be set back from the street line with areas for parking, access, turning or external storage, it is important to consider lines of defence. Defensive zones against arsonists may run parallel to lines of security and so may well be meeting dual demands: keeping out the individual intent on theft, and the intruder intent on destruction by fire.

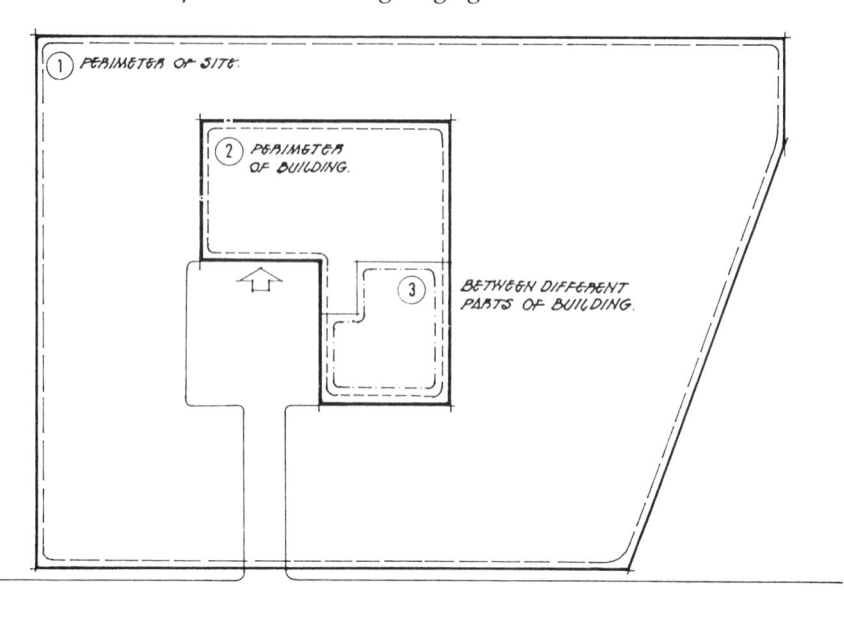

Figure 3.1 Three lines of defence.

There are three lines of defence (Figure 3.1) around a building: first, the perimeter of the site; secondly, the building face; and thirdly, the divisions between different parts within the building. Obviously the amount of damage a fire causes will increase as each of these lines of defence is breached. At the first line of defence, it is desirable to have some form of fence, hedge or barrier and permitted entrances should be marked by, at least, symbolic gateways. The design of this public-to-private interface is crucial to the success of the building resisting arson attack. There are key symbolic indicators which offer perceived zones of control. Physical barriers might be the most immediate reaction for add-on solutions, but it must be remembered that carefully designed hard landscape, including pavements, low walls, access entrances and exits, together with strong lines of sight, can do much to dissuade the intruder. Effective lighting with emergency back-up can ensure that the public-to-private zone is never in darkness, thus reducing the likelihood of unauthorized access. If the last resort is to erect tall fences with security gates and grills, the design of these can be executed in an unobtrusive manner. Screening with soft landscaping, variation of ground level or water features may all interact to conceal a sometimes inappropriate industrial-type perimeter fence. Regular maintenance and upkeep of this zone will do much to prevent the perimeter from appearing to be in a state of neglect. The costs of ongoing maintenance and repair should be borne in mind when preparing the specification. If the costs of replacing material damage are exorbitant, the occupier will decline to carry it out and consequently the effectiveness of this deterrent is lost.

At the building face, the architect wants to control the number of entrances and ensure that the grounds of the building are protected by passive surveillance. Such surveillance does not necessitate continuous observation, rather it should convey to would-be intruders the feeling that they are indeed being observed. Moving closer to the totally private zone designers should take into consideration the external face of the building, the building envelope and the factors likely to render a design vulnerable to arson attack. The management and control of access through the external envelope needs to be prioritized. With multi-use developments, there can be several categories of access required. Non-staff entrances, emergency routes, service bays and plant rooms all demand a breach of the external envelope and should be considered at the early stage of design development.

The third line of defence is within the building, and here the consideration of circulation routes is very important. Normally non-staff circulation should be kept to a minimum, and where large public spaces are inevitable, staff circulation should be planned to provide passive surveillance. Closed-circuit TV now offers an additional means of extending surveillance and, again, it is not necessary to have someone always monitoring the system for it to be effective in deterring arsonists.

3.3 THE BRIEFING, DESIGN AND CONSTRUCTION PROCESS

It is at the formulation of the briefing document that designers should be asking searching questions of the client. Often clients make rash assumptions about the architect's knowledge of their requirements. Difficulties will arise when the immediate client is not the final end-user; examples of this are speculative developments such as retail shopping centres, commercial warehouses with tenants large and small and industrial units within development zones. Recent initiatives by the public sector have sought to encourage pump-priming for the private sector by introducing new build projects in dilapidated urban areas; and by introducing enterprise zones, urban regeneration schemes and capital funding for certain categories of development. By their very existence, these new buildings can become a sensitive target for the arsonist.

In these developments design and management are integral to the success of the venture. Communication to end-users through the preparation of a management manual needs to address issues of responsibility. Who is responsible for the final fitting out of the shell? Are there gaps in overlap between building owner and building occupier which are likely to cause a failure of response, thus giving the arsonist the opportunity to attack?

Particular attention must be paid to large-scale developments that are not going to be completed and occupied in one transaction. Phased completion enables the developer to recoup capital outlay by leasing identified units ahead of others, thus allowing partial occupation while remaining units are being completed. Vacant, unattended phases are vulnerable. This is especially so when the main contractor reaches the end of his contract for the shell, removes his site security and hands over the unit to the developer. It is at this junction that the risk of arson becomes apparent. Frequently it is at this transition, too, that an insurance company is asked to insure the risk. Assessment of that risk will take into account the management factor and the measures already in place to reduce the incidence of arson.

One example of poor forward planning and ignorance of the management requirements is that of a badly designed sprinkler distributions system: to connect a new unit into the main system necessitated an entire draining down of the complete installation resulting in extra costs, disruption to business and a breach in fire prevention measures at a very critical point. Good design facilitates better management, reducing frustration by the occupier and owner. These are real and sometimes complex situations; designers could assist their clients by forward planning together with structural management for the end-user. Sometimes, clients have to be reminded of their obligations to perceive the problem and provide the necessary information to help the design team.

Having identified the external envelope as a key element against arson attack, some very simple or common fault may then occur at handover. The main contractor has during the progress of the works instructed his labourforce or subcontractors to install doors and ironmongery. Traditionally this is known as 'second fix joinery', undertaken by joiners who fitted doors and locks and handles at one and the same time. Access doors, entrance doors, security doors and service bay roller shutters are all installed with keys for locks. The keys begin to circulate to an ever widening group of people. Eventually almost everyone employed in the construction has to come and go through the envelope: keys get lost and more keys are cut – to the extent that at the point of handover everyone on site has a key, or knows someone who has a key or can obtain one. What value is external security in the envelope to the client at this stage? The arsonist is offered an opportunity by virtue of careless handling. Once again, the occupier is faced with a frustrating problem: the insurance underwriter demands that all locks are to be changed, resulting in duplication of ironmongery and abortive work, together with increased costs.

A better solution would include restriction of the issue of original keys throughout the workforce, or establishing a process by which adequate temporary security is possible until the occupier moves in. Then the final locks are fitted to the specification demanded by the insurer. In this way, selected personnel are authorized to act and are held responsible to management for their actions.

Thus limiting the number of people likely to have the opportunity to breach the external envelope, there is less opportunity for arson to take place. Within the building design against arson often correlates with fire prevention measures which have already been given consideration, in order to meet statutory requirements. One significant factor is that the determined arsonist may seek to overrule the normal fire prevention measures by means of their special intent.

3.4 TYPES OF ARSON

One insurance company has categorized arson into two types. The first covers deliberate fire-raising for gratification, revenge or racist/political ends. The arsonist in this case may not be worried about concealment of the crime. The second type can be described as fraudulent arson. In this category the arsonist is hiding a loss, attempting to claim insurance benefit by making the fire ignition look accidental or trying to conceal a crime.

In designing against the malicious arsonist the normal active fire prevention measures may be totally inadequate. In recent acts of terrorist attack, involving fires in commercial property, the arsonists deliberately set multiple seats of fire to make certain that their actions were successful. Installations of auto-suppression systems tends to centre on the single seat of fire, thus the normal systems may be insufficient for the particular terrorist act. However, Fire Prevention Officers have often argued that the normal sprinkler system does afford a minimal amount of time, and even if the water supply becomes exhausted, the Fire Brigade may be able to use the pipework systematically to tackle the blaze.

Other experts would add a third type, embracing casual vandalism, where there is less planning of the intent to cause fire, and where illegal entry to the premises is probably not initially the prime reason for fire developing. Particular examples of this type of arson are most evident in the practice of glue sniffing by schoolchildren and youths, carried out in secret away from the public eye, often within vacant or unoccupied premises during the hours of darkness. Carelessness results and a fire is started which eventually destroys the entire property and sometimes causes fatalities among the intruders. To reduce this risk securing the external envelope is important, and within the property compartmentation should be considered, if possible, even if the fire load and volume would ordinarily rule it out. The fire at the Summerland complex on the

Isle of Man in 1973 was apparently caused by young boys smoking in a shed, which when it accidentally ignited collapsed against the outer face of the leisure centre and, in turn, ignited this and eventually caused 50 deaths.

3.5 BALANCING RISKS

Achieving a balance between risk and prevention of arson necessitates an open mind on the part of the client, with a realistic assessment of the likely occurrence taken in consultation with statutory bodies, crime prevention officers and insurers. Improved communication with insurers at the design stage could result in financial savings, benefiting the building occupier who is attempting to obtain fire insurance cover at affordable rates. Architects need to inform their clients that specific forms of construction while meeting the requirements of building regulations and fire recommendations are rated inferior by some insurers. Thus the fabric of the building is downgraded, attracting a higher fire premium or no cover at all. A framework needs to be established by which these factors can be brought to the discussion table as designs are developing and specifications for construction are evolving. Several of the larger brokers have established a guidance policy serving the very large-scale developments. With guidance, the architect can assist the client with even the most modest project.

In the recent past it has been common practice for the public sector client to demand high specifications in the building envelope and internal services, both mechanical and electrical. Some private sector clients also set high standards for the construction of their buildings. In between these, however, there are private sector clients whose ideas are more ambitious than their financial resources; inevitably the architect is then obliged to reduce specification to achieve cost savings and too often the standard of passive and active fire prevention measures suffer along with the standard of construction, finishes and services. A recent trend in speculative developments has been for the developer to offload the project as soon as occupation is complete, thereby passing on responsibility for the longer-term life-cycle costs of building maintenance. If design against arson is going to be successful, it should not be downgraded in priority.

In projects where the eventual building use is identifiable, the architect has an opportunity to plan for a reduced risk of arson taking place within the interior. Commercial warehouse design with high-value, high fire-load contents need special forms of construction for safe storage. Examples include giving adequate priority to consideration of the management, movement and provision of safe storage for cigarettes, electrical and electronic goods, flammable liquids and gases. Economic trends and patterns of taxation can greatly increase the throughput of tobacco goods, particularly at the time of the government's Budget, with a hint of imminent changes in taxation or legislation. The storage of tobacco goods by a distributor within a warehouse may considerably increase in volume over a short period of time. Designated storage areas for sensitive goods are not of much assistance if they are too small or unusable at peak throughput times. The design team can play a role in planning for long- and short-term storage which is adequately fire rated in construction within a larger area, thus rendering the goods less readily accessible to the arsonists.

In buildings housing production processes, such as factories or assembly plants with large spaces and open areas, the careful design of circulation routes can be a deterrent to the would-be arsonist. Patterns of occupations, together with the actual type of operational workload, can result in boredom for the employee; unsupervised, lengthy periods of repetitive work could lead to arson as a ploy, just to get attention, obtain kicks or alleviate the boredom. Sensitive design of circulation routes can prevent personnel from feeling isolated or under-supervised. The movement of people through the workplace on the factory floor or overhead will provide management with a better perception of its workforce and a more immediate visual response where problems occur. The interrelationship between circulation and passive surveillance should be highlighted in the design process by exploring the relationship between employees, management and the proposed operation of the building.

Where the building use has high throughput of waste, the design brief should consider adequate provision for storage space; often this space is viewed as non-productive or non-profit making, and as a consequence little consideration is given to its design. This may give the arsonist a window of opportunity to ignite large volumes of flammable waste which are carelessly stored. If the nature of the waste demands that it is stored internally until disposal, then the design teams need to take this into account, plan accordingly with suitable construction of appropriate fire rating. Should the waste be stored externally, then provision must be made within the external zone to ensure that it is not piled up against the building envelope and not close to the perimeter to make ignition an easy goal for the arsonist (Figure 3.2)

Where the production process within the building results in high-value goods or high-value intelligence, the designer should assist the client by designing to mitigate against the consequential loss factor. Business interruptions can be more of a financial burden than the actual cost of fire loss of the building. Consideration should be given to compartmentation of the storage of these goods and, in some cases, off-site duplication of intelligence storage and systems are appropriate.

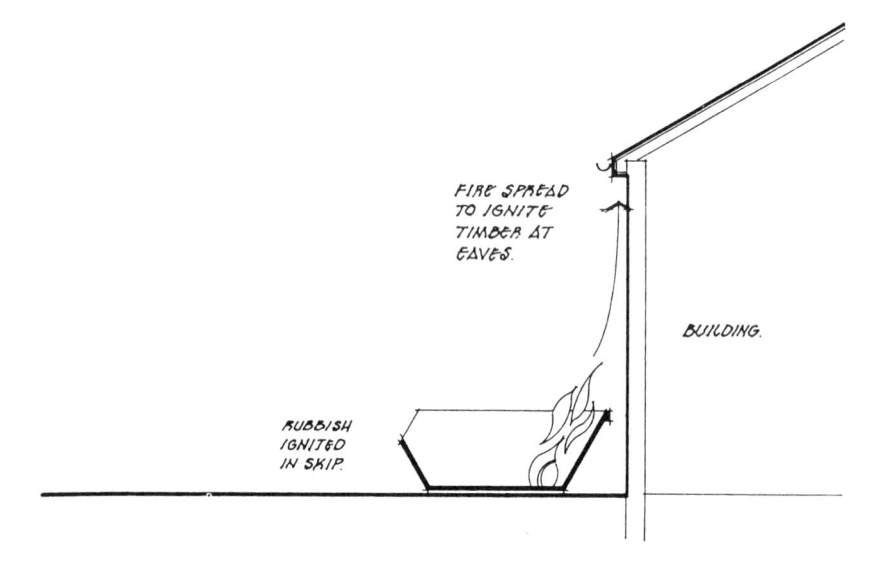

Figure 3.2 Skip fires spreading to eaves.

In designing against arson it is not sufficient to rely only on fire prevention; the other fire safety tactics of communications, escape, containment and extinguishment are all critical in mitigating fires which are ignited.

ACKNOWLEDGEMENTS

The advice and assistance of Peter Bayliss (Royal Insurance), Charles Browne (Association of British Insurers) and David McMinn (Legal and General Insurance) is gratefully acknowledged.

REFERENCES

1. Association of British Insurers, press release, 15 February 1991.
2. Home Office (1988) *The Prevention of Arson*, Report, HMSO, London, December.
3. Department of Education and Science (1988) *Fire and the Design of Educational Buildings*, Building Bulletin No. 7, HMSO, London.

4

Fire communications

John Northey

4.1 INTRODUCTION

The objective of a fire alarm system is to give early, reliable warning in case of fire such that action can be taken to avoid life loss and to keep property losses to a minimum. Its design is inevitably a compromise between sensitivity to fire, its propensity to give false alarm and the available technology and its cost. Fire alarm systems range from simple manual call point systems to complex analogue addressable systems; Figure 4.1 shows the fundamental components which are found in a fire detection system.

A fire communications system must contain a detection device (an automatic detector or manual call point), a control and indicating panel, a power source and a means of giving warning in case of fire (usually an audible warning sounder). Within the system it is essential that all these components are compatible with one another; care needs to be taken to select the right type of detector for the particular application, and there is a need for understanding of what the various systems have to offer. Advice on system design, installation and servicing is given in the relevant part of British Standard 5839 [1,2,4], which covers all fire alarm systems from modern microprocessor-based addressable analogue systems to simple manual call point systems. This standard can be used as a specifying document and will be referred to throughout this chapter.

4.2 CHOOSING THE DETECTOR

The three fire characteristics most commonly used for fire detection are heat, smoke and radiation from a flame; Figure 4.2 indicates the wide variety of detectors available. All these detectors have both advantages and disadvantages, no one type of detector is the most suitable for all applications. In choosing the most suitable detector, or detectors, account must be taken of the likely fire behaviour of the building's contents, the

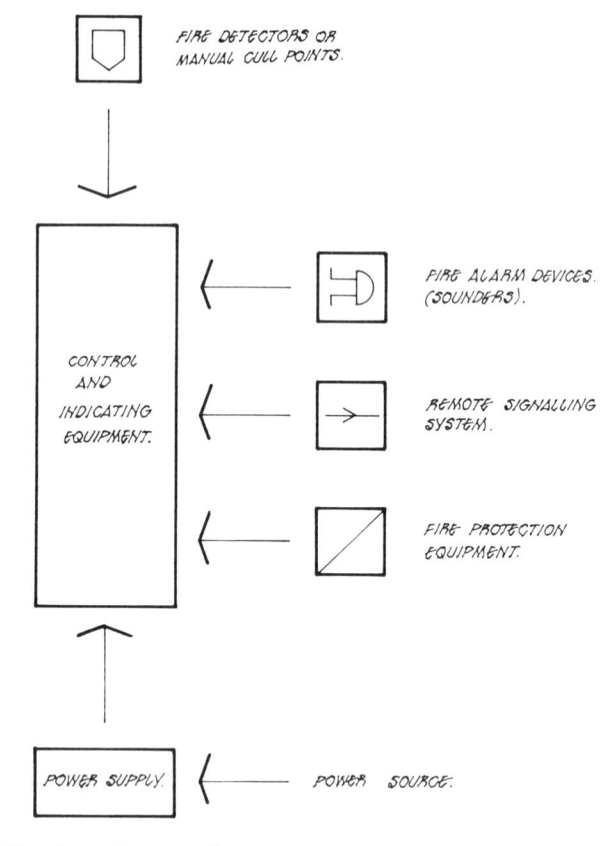

Figure 4.1 Fire detection and alarm system.

processes taking place, the design of the building and the primary purpose of the system – that is, is it intended mainly for life safety or property protection? Heat and smoke detectors will be suitable for most buildings. Generally smoke detectors will detect fires quicker than heat detectors, but they are more susceptible to false alarms.

A factor which cannot be ignored is the height of the ceiling area being covered. Point detectors are generally only suitable for use in areas where the ceiling height is below 15 m. Table 4.1 [1] gives an indication of the ceiling height limits for heat and smoke detectors. From this table it can be seen that long-range optical beam detectors are to be used in buildings with ceiling heights up to 40 m, provided that there is a link between the building's fire detection system and the Fire Brigade, and that the Fire Brigade can arrive on the scene within 5 min. Where a rapid attendance by the Fire Brigade is not achievable, then the ceiling height limit is 25 m.

In tall buildings flame detectors can be used, either alone or to supplement heat and smoke detectors. The flame detectors will need to

Heat detectors

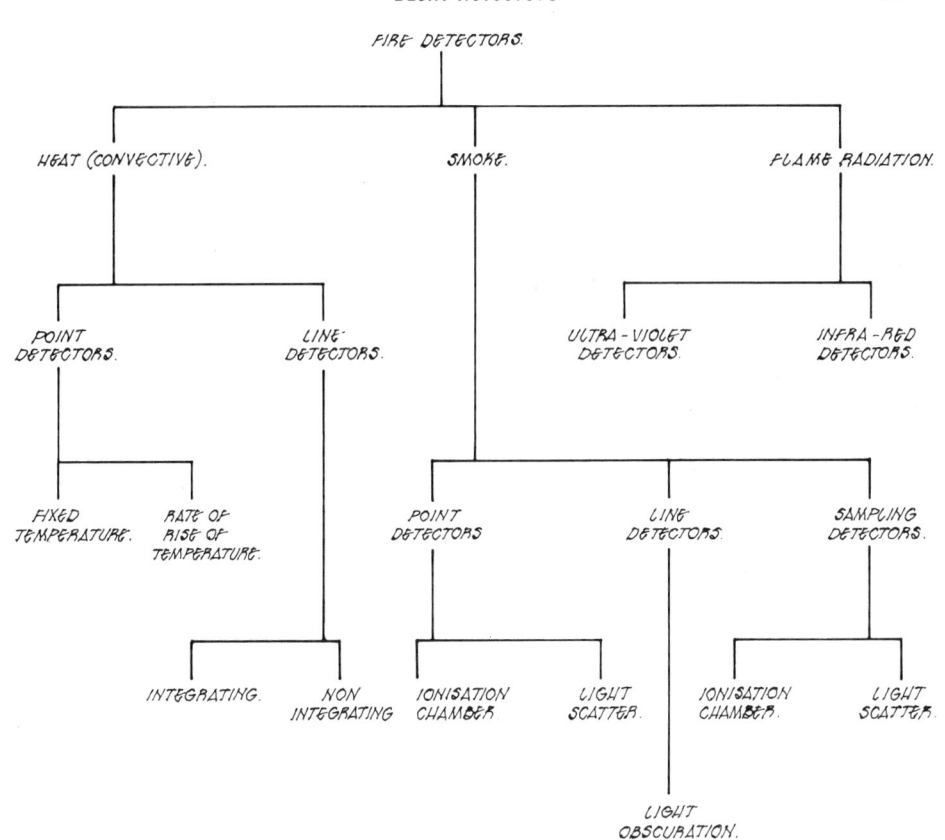

Figure 4.2 Types of fire detectors.

be sited so as to have an unobstructed view of the area they are protecting; flame detectors also offer the best solution for outdoor applications. Another application in which flame detectors are likely to be the most suitable is in areas where there is a danger of flammable liquid fire as they respond to flaming fires extremely quickly.

The foregoing considerations mainly are general ones. Before moving on to consider system features, a few detailed points relating to the available detectors and some specific fire detection applications are worth noting.

4.3 HEAT DETECTORS

There are two main categories of heat detectors: the point type and line type. Point-type detectors respond to the temperature of gases in the immediate vicinity of a single point, and line-type detectors respond to the temperature in the vicinity of a line (not necessarily straight).

Table 4.1 Limits on ceiling heights (BS 5839, Part 1: 1988)

	Ceiling heights	
	General (m)	Rapid attendance (m)
Heat detectors (BS 5445: Part 5)		
grade 1	9.0	13.5
2	7.5	12.0
3	6.0	10.5
Point smoke detectors	10.5	15.0
High-temperature heat detectors (BS 5445: Part 8)	6.0	10.5
Optical beam smoke detectors (BS 5839: part 5)	25.0	40.0

Note: To qualify for 'rapid attendance' the system has to be automatically connected to the Fire Brigade directly or via a central (fire alarm) station and the usual attendance time of the Fire Brigade has to be not more than 5 min.

Line-type detectors are either integrating or non-integrating. The alarm temperature for integrating detectors depends on the length of line heated. In the case of non-integrating detectors, the alarm temperature is independent of the length of line heated. Another variation is that they can either be of the recoverable or non-recoverable type. The non-recoverable type are effectively destroyed when the alarm temperature is reached; in one such type, the alarm is activated by conducting wires making contact when the insulation between them is destroyed at a particular temperature. In the recoverable type the sensing element restores to a fully operational state upon a reduction in temperature below the alarm threshold. The advantages of the recoverable over the non-recoverable type are obvious in terms of replacement needs. The recoverable type can also offer, with the appropriate control equipment, the advantages which can be gained by an analogue system such as an adjustable alarm threshold and pre-alarm warnings.

Line detectors tend to be used primarily for specific detection applications, whereas point detectors are normally used in general fire detection systems in buildings. The types of application for which line detectors are used include protecting cable trays, escalators, cold storage warehouses, power generation plant, aircraft engines and military vehicles.

There are currently no international (ISO) or European standards for line heat detectors, but there are three European standards for heat-sensitive point detectors. EN54: Part 5 covers heat-sensitive point detectors containing a static element; EN54: Part 6 covers rate of rise point detectors without a static element; and EN54: Part 8 covers high-temperature heat detectors. All three standards are currently being revised and an attempt made to incorporate them into one standard. EN54: Part 6 was not accepted as a British Standard. The UK voted against the standard as it dealt purely with a rate of rise heat detector; the UK believes that a heat detector should respond if the temperature reaches a sufficiently high level, even if the temperature rise is very slow.

4.4 SMOKE DETECTORS

There are two main types of smoke detectors: ionization chamber smoke detectors and optical smoke detectors. Ionization chamber smoke detectors are basically small particle detectors, and in fact they detect small particles which are invisible to the naked eye; they operate due to the fact that the electric current flowing between electrodes in the detecting ionization chamber is reduced when smoke particles enter the chamber. Optical smoke detectors detect large particles by detecting the scattering or absorption of light by particles.

Ionization detectors are very suitable for detecting smoke particles from cleaning fires but may be less sensitive than optical detectors to the large particles produced by smouldering fires. Optical detectors are less sensitive to small smoke particles. Generally, though, both types of smoke detector have a sufficiently wide range of response to be of general use; however, in some premises there may be specific risks where one type may be more suitable than the other.

There is a European standard for point-type smoke detectors, EN54: Part 7, which is currently being revised, and a European standard for optical beam detectors is being prepared. In addition, international standards (ISO) for point-type smoke detectors and smoke alarms for use in single family dwellings are being prepared.

4.5 SAMPLING/ASPIRATING DETECTORS

The basis of an aspirating system is the use of a tube and fan to move air from the protected area to a remote detection point. In some systems, the tube feeding the detector will have several inlet points; in other systems, the air is drawn from several tubes which are sampled sequentially by the use of a rotating valve arrangement. Where there are several input points, dilution will occur if smoke is entering one and clean air entering the rest, thus on a one-to-one basis aspirating systems may be less sensitive than point or beam detectors. In many modern systems, though, compensation is made for the diluting effect by using a detector which is much more sensitive than a point detector.

Applications in which aspirating systems may have an advantage over other detection systems include: the protection of historic buildings, where the presence of point-type detectors may be considered unsightly; air conditioned rooms or equipment, such as computer cabinets, where the aspirating system can be designed to cope with the dilution of smoke caused by air movement; high ceiling areas, where the aspirating system can be made to be very sensitive so as to overcome the dilution of the smoke plume; and cold stores, where the tubing in the cold store needs to be plastic to discourage icing and have an internal diameter of not less than 25 mm and with the aspirating detector positioned outside the cold store with at least 10 m of tubing in a warm environment. There are no international, European or British Standards for aspirating systems.

4.6 FLAME DETECTORS

Flame detectors, either detect ultraviolet radiation or infra-red radiation, employ photoelectric cells which 'see' the fire directly or respond when radiation is reflected upon them. (A European standard for flame detectors is in preparation, but not an international standard.)

4.7 MANUAL CALL POINTS

Manual call points must be easily identifiable and simple to use. It is desirable that in an installation the method of operation of all manual call points should be identical. The British Standards Code of Practice for manual call points [2] states that a person operating a manual call point should not be left in any doubt as to the success of the operation; it advises that the delay between the operation of a call point and the giving of the general alarm should not exceed 3 s.

The Code requires manual call points to have a frangible element, usually glass, which must be broken in order to operate the call point. This element has to be replaced to make the call point operational again. The reason for using this method of operation is to deter nuisance. Another type which is available does not include a truly frangible element; the element appears to break on operating the call point in much the same way as a frangible element, but in fact it can be re-set and is unlikely ever to need replacement. These latter type of call points do not meet the current British Standard but are recognized in the draft European standard. They may be particularly suitable for use in, for example, food preparation areas, although call points meeting the requirements of the Code are also available which do not release fragments of glass when they are operated. In one example, a plastic film prevents fragments of glass being released when the call point is operated. Another area where special care needs to be taken is in the selection of the appropriate manual call point where a flammable or explosive atmosphere may be present; and here there are special requirements.

Manual call points should be located on exit routes and, in particular, on floor landings and stairways and at exits to the open air. It is recommended that they should be sited such that no person has to travel more than 30 m to reach one and in special cases, for example, where the occupants are likely to be slow in movement or where potentially hazardous conditions exist, they may need to be sited at closer distances. The height at which they are to be fixed should not be over 1.4 m from the floor. Call points must be readily accessible, conspicuous and in an illuminated area.

4.8 WARNING DEVICES

A wide variety of sounders are available on the market and care should be taken to ensure that all fire alarm sounders within a building have similar sound characteristics, unless particular conditions makes this impracticable. The note of the fire alarm sounders should be distinct from any other sounder likely to be heard in the building and, in particular, it should be distinct from the fault warning signal given in the control equipment.

In the British Standards Code of Practice [1] advice is given on the sound levels required, the use of two-stage alarms and the frequency of the sound; the recommendations are:

1. Audibility – in all accessible parts of the building a minimum sound level of 65 dB(A), or 5 dB(A) over any background noise likely to persist for a period of longer than 30 s, and whichever is the greater, is required, and if sleeping persons need to be woken, the minimum sound level should be 75 dB(A) at the bedhead with all doors shut.
2. Sound patterns – where a two-stage alarm is required, the alert signal should be intermittent, 0.5–1.5 s on and 0.5–1.5 s off, and the evacuate signal should be a continuous or warble tone.
3. Frequency – sounder frequencies should lie between 500 Hz and 1000 Hz as this provides a good combination between the sensitivity of the human ear and the ability of the sound to penetrate the fabric of buildings (doors, partitions, etc.).

An important point referred to in the Code is that the removal of any fire detectors in the system from their bases should not affect the operation of the sounders.

There should be at least one alarm sounder in every fire compartment; the Code also points out that a larger number of quieter sounders, rather than a few very loud sounders, may be preferable in order to prevent excessive sound levels in some areas.

In some buildings, particularly large ones, consideration should be given to using public address equipment to give warning in case of fire in lieu of conventional sounders – voice alarm signals are likely to offer major benefits over use of sounders alone. If a voice alarm system is to be installed, then reference should be made to the relevant British Standard [3], which is technically equivalent to IEC 849: 1989, as well as the Code for alarm and detection systems [1]. Some of the recommendations for the use of voice alarm equipment are as follows:

1. The voice alarm equipment's power supply and standby duration should comply with that required for a fire detection system.
2. Amplifiers should be built to the relevant clauses of British Standard [4] covering construction requirements and electrical requirements.
3. Audibility should be similar to that specified for tone sounders.
4. Only the speech modules (or equivalent message generators) which give the warning and those microphones designated as fire microphones should be operational in an alarm and they should override all other duties (the system should be designed such that it is not possible for messages from more than one microphone, speech module or message generator to be broadcast simultaneously).
5. All automatic announcements should be retrieved from a non-mechanical memory.
6. Messages should be short and unambiguous. Intervals between messages should not exceed 30 s and periods of silence should not exceed 10 s.

4.9 CONTROL EQUIPMENT AND SYSTEMS

In any system the detectors must be electronically compatible with the control and indicating equipment, otherwise the system will not function. Control and indicating equipment panels range from those for a single system to those for addressable analogue detection systems.

The simple systems (Figure 4.3), in which the detectors are designed to send a signal to the control panel informing it of a fire when a certain pre-set limit has been reached and the panel then signals the fire and triggers the alarm bells, are generally cheaper than the more complex systems. Some savings are achievable in wiring, though, when an addressable system (Figure 4.4) is installed because the detectors in a number of zones are wired in a single wiring loop, and above a certain size of system, this could result in an addressable system being less expensive than a simple zoning system.

Control equipment and systems 49

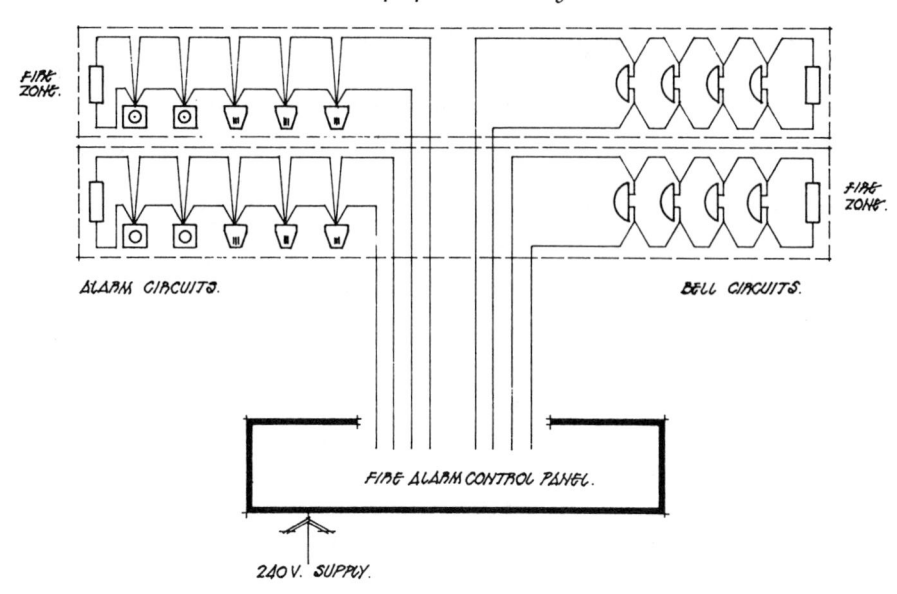

Figure 4.3 Conventional fire alarm system.

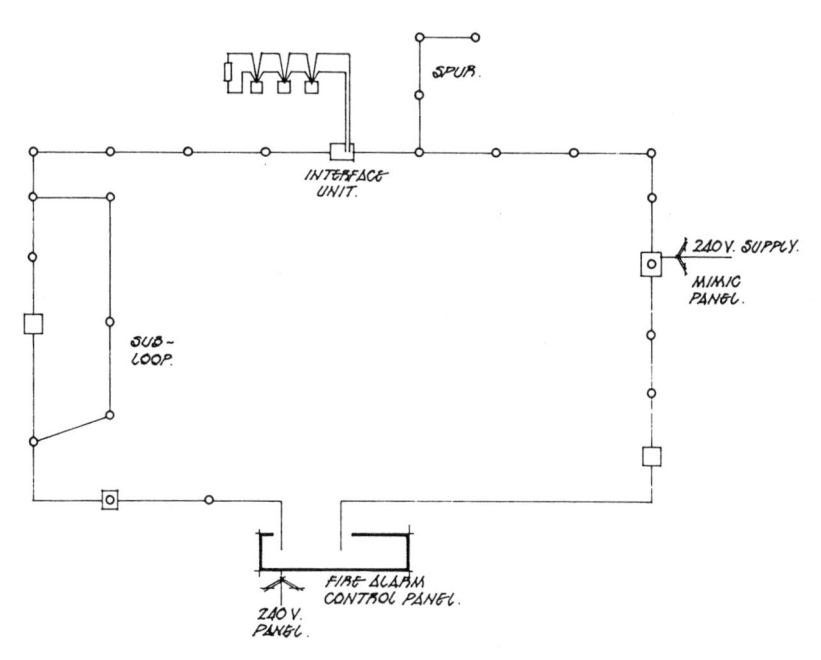

Figure 4.4 Addressable fire alarm system.

In an analogue system the detectors send information to the control panel, and it is at the panel that the information is processed; some of the advantages claimed for analogue addressable systems are:

1. A reduction in false alarms (this arises from the system's ability to compensate automatically for drifts in detector sensitivity).
2. Identification of detectors (this is valuable from the point of view of locating a fire or a faulty detector).
3. Cable economics.
4. The sensitivity of detectors can be programmed in a time-related manner (i.e. they can be made more sensitive during certain times than others, thus reducing the possibility of false alarms without compromising on sensitivity when it is needed).

Analogue addressable systems can be interfaced with conventional systems, and although many – if not most – of them are hard wired, they can utilize radio interconnected detectors which report back to a control panel. A radio interconnected fire alarm system is just another of the choices available. Like a hard-wired system, it has advantages and disadvantages.

The European Standards Committee (CEN) is preparing a standard for fire detection and fire alarm systems, control and indicating equipment, and power supplies. It is also preparing an installation standard for detection and fire alarm systems. The British Standard for control and indicating equipment is BS 5839: Part 4(4).

4.10 INSTALLATION REQUIREMENTS

The British Standards Code of Practice for alarm and detection systems [1] specifies the installation requirements for different uses for which the fire alarm system may be required. It gives recommendations on the spacing of detectors, wiring requirements, power supplies, maintenance and commissioning, plus a number of other important factors which need to be considered when designing and installing a fire alarm system. The different uses for a fire detection and alarm system specified are:

- Type P systems are automatic detection systems intended for the protection of property; they are further subdivided into:
    P1 – systems installed throughout the protected building;
    P2 – systems installed only in defined parts of the protected building.
- Type L systems are automatic detection systems intended for the protection of life; they are further subdivided into:
    L1 – systems installed throughout the protected building;
    L2 – systems installed only in defined parts of the protected building (type L2 system should normally include the coverage required of a type L3 system);

L3 – systems installed only for the protection of escape routes (note that to give satisfactory protection of escape routes detectors should be installed both on escape routes and in rooms opening onto escape routes).
- Type M systems are manual alarm systems and have no further subdivision.

Systems intended for use in multi-occupancy buildings are given the suffix letter 'x'.

It is advised that premises protected by automatic detection systems, both of types L and P, should normally also be provided with manual call points.

If a system is intended to fulfil the purpose of more than one type of system, and the recommendations for the types differ, then the system should comply with the recommendations for each of the types. This should not cause problems, and although there are detailed differences in the installation requirements of the systems designed for various uses, many features are common to all fire alarm systems. The differences are likely to be in the number and type of detectors used and the back-up power supply requirements.

With regard to the positioning of heat and smoke detectors, account has to be taken of the convective movement of fire products from the fire to the detector. The spacing and siting of detectors is based on the need to restrict the time taken for this movement and to ensure that the products reach the detector in adequate concentration. As a rough guide, the number of point-type detectors fitted in any room or compartment should not normally be less than one per 100 m^2 for smoke detectors, or less than one per 50 m^2 for heat detectors.

In the case of beam detectors, the manufacturer's instructions must be followed. The Code advises that because of the uncertainty in the position of a fire within the beam length, the maximum length of the area protected by a single optical beam should not exceed 100 m. If there is a possibility of people moving about in the area of the beam, then the beam should be at least 2.7 m above the floor.

After the detectors, the most important matter to consider is the interconnections between the components of the system. Most connections, other than those to detectors or call points, will be required to function correctly for significant periods during a fire. Such connections include those by which power is supplied to the control equipment, and those linking the control equipment to the alarm sounders. In general, it may be assumed that interconnections between sounders, control and indicating equipment and power supplies which can resist fire for at least half an hour will be satisfactory. In special cases, a longer period may be required; for example, in staged alarm systems. It is recommended that conductors carrying fire alarm power or signals should be separated from conductors used for other systems.

There are limitations on the effect that one or more wiring faults can be allowed to have on a system:

1. If separate circuits are used for each zone, then a fault, or faults on one circuit cannot affect any other circuit.
2. If any circuit is used for more than one zone, then a single fault on that circuit cannot remove protection from an area greater than that allowed for a single zone.
3. If a circuit is used for more than one zone and multiple faults within one fire compartment could remove protection from an area greater than that allowed for a zone, then the circuit within that compartment is suitably protected.
4. If two simultaneous faults occur, they should not remove protection from an area greater than $10\,000\,m^2$.

Except in small manual systems, wiring is required to be monitored. Furthermore, if an open circuit or a short circuit occurs, a fault warning must be given and not an alarm of fire.

Fault warning condition responses, in respect of wired systems, are required to be given within 100 s of the fault occurring if the fault disables one or more detectors or call points; if not, a fault should be indicated within 60 min. In the case of radio systems, because limitations of allowed frequency spectrum can lead to interference between simultaneous signals, it is unwise to send monitoring signals at very frequent intervals. Radio links should be monitored in such a way that if signals are not received from a detector, call point, sounder or other remote component, then the failure will be indicated at the central control and indicating equipment within a period of 4.5 h from the occurrence of the fault.

4.11 INDICATING THE FIRE

A satisfactory fire alarm system is one that automatically detects a fire at an early stage, raises an effective alarm and indicates the location of the fire in the building. With modern systems, it is possible to indicate precisely which detector has operated, but the primary indication of the origin of the alarm should be an indication of the zone of origin. A display identifying only individual detectors may be difficult to interpret and could lead to difficulties in assessing the spread of fire or the occurrence of secondary fires. Any display of individual detectors should be subsidiary to the zone display.

The floor area of a single zone should not exceed $2000\,m^2$ and the search distance, or distance that has to be travelled by a searcher inside the zone in order to determine visually the position of the fire, should not exceed

30 m. Zone boundaries, where they extend beyond a single compartment, should be boundaries of fire compartments. It is thus permissible to have two complete fire compartments in one zone, or two complete zones in one fire compartment. It is not permissible to have a zone which extends into parts of two compartments, or a compartment which extends into parts of two zones. Apart from small buildings, that is those having a floor area not greater than 300 m^2, zones should be restricted to a single storey, except in the case of stairwells, lightwells and other flue-like structures which are within one fire compartment.

For a fire detection system to give maximum benefit, its alarm needs to be passed on to the Fire Brigade with the least possible delay. If there are adequately trained people on the premises, then this may be achieved by using the telephone. Frequently, however, the only reliable method will be an automatic link to a commercial central fire alarm station which, in turn, will pass the call on to the Fire Brigade.

4.12 USERS' RESPONSIBILITIES

Attention to a fire alarm system does not stop after its design and installation, for there is a need to ensure it is properly maintained and serviced. It is recommended that normally an agreement should be made with a manufacturer, supplier or other competent contractor for regular servicing.

There is also a need to appoint a 'responsible person' to control the fire alarm system, check it and ensure that it is serviced. Scattered throughout the British Standards Code of Practice [1] are recommendations as to the duties this person should be allocated. In Peter Burry's book [5] a list of these duties is given, and this is reproduced below.

- In conjunction with the appropriate fire authority (usually the local Fire Brigade), to lay down procedures appropriate to the premises for dealing with the various alarms, warnings or other events originating from the system.
- To ensure that all those who will have to use the system are instructed in its use. In particular, anyone who might be concerned with first aid fire-fighting should be trained in translating the system's indications into the position of a fire in the building. Presumably this would also apply to people who might have to control or assist in the evacuation of the building.
- To liaise with others to ensure that work on the building, such as decoration or cleaning, does not adversely affect the system, and that possible effects on the system are taken into account when planning changes to the building.
- To ensure the efficiency of the system is not reduced by obstructions

which might prevent the movement of fire products to the detector, or obscure or block access to manual call points.
- To ensure that all necessary documentation has been completed before handover, and to maintain drawings and operating instructions in suitable condition.
- To keep the logbook and ensure that entries are made including brief details of every significant event affecting or resulting from the system. At the head of the list, of course, will be the name of the 'responsible person'.
- To prevent (or reduce the rate of) false alarms.
- To ensure that the system is properly reinstated at the conclusion of any work.
- To ensure that the system is given correct routine attention at the proper intervals.
- To ensure that the system is correctly serviced following any alarm or warning it might give, and is correctly repaired following any damage it might receive.
- To maintain a stock of suitable spares, usually following agreement with a servicing organization.
- To authorize personnel to alter local information such as zone names or sensor locations.

The installer should always ensure that the 'responsible person' is aware of the duties outlined above, and of ways in which these duties are applicable to the system. Remember if the 'responsible person' has not been appointed, then the Code states that the person having control of the premises automatically takes on this role. Although the 'responsible person' is indeed held responsible, he does not have to fulfil all the duties himself; he can delegate the job, either to someone within his own organization or to an outside organization (such as a servicing company). It will be unusual for the 'responsible person' to have all the expertise required to carry out all the duties; he should be encouraged to recognize any limitations and to fill the gaps by appointing other suitably expert individuals or organizations – however, he will remain responsible for seeing that the job gets done.

A fire alarm system (although it cannot reduce the incidence of fire or automatically extinguish a fire) can help to lessen the resultant loss by reducing delay between ignition and the start of effective fire-fighting, and it can provide sufficient warning to people present in a building to enable them to escape the threatened area.

REFERENCES

1. British Standards Institution (1988) *Fire Detection and Alarm Systems in Buildings, BS 5839, Part 1: Code of Practice for System Design, Installation and Servicing*, BSI, London.

2. British Standards Institution (1983) *Fire Detection and Alarm Systems in Buildings, BS 5839, Part 2: Specification for Manual Call Points*, BSI, London.
3. British Standards Institution (1991) *Sound Systems for Emergency Purposes, BS 7443*, BSI, London.
4. British Standards Institution (1988) *Fire Detection and Alarm Systems in Buildings, BS 5839, Part 4: Specification for Control and Indicating Equipment*, 1988, BSI, London.
5. Burry, P. (1990) *Fire Detection and Alarm Systems: A Guide to the BS Code BS 5839, Part 1*, Paramount, Borehamwood.

5
Escape behaviour in fires and evacuations

Jonathan Sime

5.1 INTRODUCTION

The aim of this chapter is to illustrate the way in which knowledge of the principles and the findings of research on people's escape behaviour in fires and evacuations are essential in fire safety engineering design. The first part of the chapter is a review of research on escape behaviour in fires and evacuations in public settings; the second part summarizes four illustrative research studies conducted by the author and colleagues. In seeking to inform design it is considered that there is insufficient emphasis on research of human behaviour in fire safety codes, design education, the practice of architectural and engineering design and indeed, to date, in the evolving disciplines of fire science and fire safety engineering [1].

While some fire safety design textbooks [2,3] implicitly consider human behaviour in fires and evacuations, explicit reference to relevant research over the past 20 years is limited to only a few paragraphs and references. The predominant emphasis is on the nature of fire and smoke spread, fire protection and risk assessment under headings such as 'Designing for fire safety' and 'Buildings and fire', not the relationship between people, buildings and fires. In listing areas of fire engineering consultant expertise outside the architect's normal competence, and which client and architect might wish to draw upon [2], no reference is made to human factors. Yet research on the way in which buildings are used and comprehended, and therefore how people behave in an emergency, is arguably at least as important to the achievement of life safety as research into smoke and fire spread, buildings, structures or fire protection systems. To achieve adequate life safety it is important for architects, engineers, Building Control Officers, Fire Protection Officers and managers of public buildings to have an adequate knowledge of research on

$A \times B \times C \times D$

human behaviour in buildings and complex public settings (e.g. football stadia, underground stations, etc.). In order that decisions by these and other professional groups are informed by relevant research, it is also essential that fire safety engineering design encompasses this knowledge.

The growing awareness of a need for inclusion of such a research knowledge base in fire safety design is reflected in the following:

- Inclusion of Part C, 'People and Fire', as one of several principal topics in the UK National Core Curriculum in Fire Safety Studies by Design [4].
- The increasing emphasis on crowd safety management in design codes for 'complex' spaces as a result of incidents such as the Bradford football stadium fire (1985) and King's Cross underground station fire (1987) [5].
- The change in the title of the British Standard BS 5588 series from 'Fire Precautions in the Design and Construction of Buildings' to 'Fire Precautions in the Design, Construction and Use of Buildings' [5,6].
- The increasing inclusion of psychological factors in fire safety engineering performance criteria applied to complex settings, in achieving acceptable and equivalent levels of safety between different forms and combinations of fire detection and warning, means of escape design, fire protection systems and evacuation management.

5.2 DESIGN × INFORMATION TECHNOLOGY × MANAGEMENT × BUILDINGS IN USE

As a guide to the human factors which should be considered in fire safety engineering design, Figure 5.1 is presented. This figure was originally developed in relation to crowd safety design, Environmental Design Evaluation (EDE), Post-Occupancy Evaluation (POE) and Post-Disaster Evaluation (PDE) in disasters such as the King's Cross fire and the Hillsborough Stadium crowd crush in 1989 [7]. In this respect, there is an overlap between fire and crowd safety. The relationship between Design (A) × Information Technology (B) × Management (C) × Buildings in Use (D) in Figure 5.1 is equivalent to the arrow between Communication and Escape in Figure 2.1, which also includes methods of fire prevention, containment and extinguishment. The term user-oriented architecture (UOA), which is given in the subtitle of Figure 5.1, indicates that in order to achieve a reasonable degree of safety in the design of a setting, the perspective of the public is paramount. The conventional approach to design is to concentrate on (A) (e.g. the architectural form), and secondarily, or at a later stage, on (B), (C) and (D). It is argued here that the order of priorities in Design should be reversed – i.e. (D) × (C) × (B) × (A). Clearly the process of designing the interactive relationship between

58 *Escape behaviour in fires and evacuations*

```
┌─────────────────┐
│      (A)        │◄──┐
│    DESIGN       │◄─┐│
└────────┬────────┘  ││
         ▼           ││
┌─────────────────┐  ││
│      (B)        │  ││
│  INFORMATION    │──┘│
│   TECHNOLOGY    │   │
└────────┬────────┘   │
         ▼            │
┌─────────────────┐   │
│      (C)        │◄──┤
│   MANAGEMENT    │   │
└────────┬────────┘   │
         ▼            │
┌─────────────────┐   │
│      (D)        │   │
│    BUILDINGS    │───┘
│     IN USE      │
└─────────────────┘
```

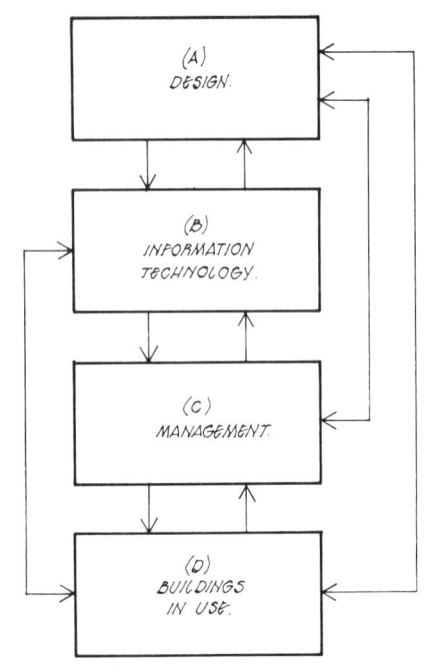

Figure 5.1 Environmental design evaluation (EDE) factors and links relevant to user-oriented architecture (UOA) for public buildings.

(A) (B) (C) (D), or (D) (C) (B) (A), is iterative and cyclical rather than strictly sequential, as indicated by the arrows. When people start to use a setting, a POE of (D) × (C) × (B) × (A) through monitored evacuation tests should also feedback or 'forward' to future designs, hence the idea of a POE design cycle and cumulative design knowledge [8].

Many of the problems for people in fires arise out of an inadequate appraisal of the way in which buildings and settings are managed, used and comprehended by people in normal circumstances. An example of this is POE research on wayfinding over the past 10 years [9,10] which has revealed ways in which certain architectural forms can be seriously disorientating (psychologically and spatially 'complex') for people in the everyday use of a public setting. The 'non-correspondence' between the outside form of a building and internal layout, for example, may compound difficulties people experience in escaping from a fire [9].

The rationale of Figure 5.1 is that appropriate 'escape behaviour design' requires an understanding of:

1. the normal use of a building (e.g. which routes are most familiar?);
2. the fact that a building or setting is not only a physical structure, but an information and communication system.

The latter is related as much to (B) Information Technology, such as alarm sirens, electronic visual displays and public address systems, as (A), the spatial layout and design of features, such as 'landmarks' which help people to orientate themselves in space. In environmental psychology, and building use and safety research [11], 'environmental cognition' refers to the fact that environments provide information; information is selectively attended to by people and anticipatory knowledge and expectations about spatial layouts and dimensions influence the environmental cues people attend to prior to and during movement [12].

5.3 MODELS OF HUMAN MOVEMENT AND BEHAVIOUR IN EMERGENCIES

There are two very different ways to 'model' (predict, describe and explain) people moving around buildings. These are in terms of movement (model A) or behaviour (model B). Model A has been called the engineering, physical science or 'ball-bearing' model of human movement [13]. This is the predominant model of escape behaviour in fire codes, fire science and engineering and media coverage of fires. The 'physical science' model of human reactions is complemented by the assumption that in extreme situations of potential entrapment people panic, characterized by the non-social or asocial behaviour of a homogeneous crowd of individuals in flight, competing for diminishing access to an exit where crushing occurs. People are assumed to be behaving irrationally, an assumption that is used to justify the idea that people's movement can be validly modelled in terms of non-thinking objects propelled to an exit and along escape routes. The emphasis in warning procedures is on keeping information about a threat away from people until it is absolutely necessary to 'avoid panic' [14]. Model A assumes environmental determinism. According to the logic of model A, the time for people to escape and the direction of movement is determined by the speed of fire and smoke spread, the numbers of people present and dimensions of escape routes: proximity of and travel distances to exits and width of exits (criteria used to define means of escape in fire design codes).

Model B, is a 'social science' or psychological model of human reactions, which considers people to be 'active agents' who, unlike ball-bearings, think and act according to the available information, social factors such as group ties and their role within a building (i.e. staff or public). The emphasis is on patterns of behaviour (actions, goals and cognition), rather than where this behaviour takes place. The sociological or psychological model, as represented by research of disasters over the past few decades, puts an emphasis on representing and explaining people's behaviour from the perspective of the people escaping, in contrast to other people's interpretation of the behaviour. In this respect, flight behaviour is usually deemed to be rational from the perspective of people warned too late to evacuate a setting in an orderly fashion. Flight is paradoxically the self-fulfilling consequence of a delay in warning. While model B emphasizes the information people need to act appropriately (something model A hardly considers), it has been relatively poor at defining the information and patterns of behaviour in terms of the physical environment. Thus both models have their strengths and weaknesses and need to be reconciled. According to the logic of model B, the time for people to escape is determined predominantly by the information available to them about a fire threat – i.e. the time it takes people to start to move, knowledge of a setting, social constraints, visual access to exits, familiarity with escape routes and wayfinding.

5.4 RESEARCH ON ESCAPE BEHAVIOUR

Research of escape movement and behaviour can be characterized under four headings:

1. Laboratory experiments.
2. Evacuation research.
3. Computer simulations.
4. Questionnaire surveys and case studies through interviews.

The range of research strategies adopted is a reflection of the practical and ethical problems of researching behaviour in fires, the models espoused and focus of the studies on people or the environment. The research strategies, broadly speaking, increase from laboratory experiments to case studies in terms of validity, the degree to which the phenomenon of human behaviour in fires has been studied directly, and decrease in terms of reliability, or accuracy with which the factors which could influence behaviour are controlled, monitored and can be reproduced. There are of course exceptions to this. A well-designed and conducted experiment can have greater validity than a poorly designed questionnaire survey. However, a range of laboratory experiments have equated the behaviour of individuals avoiding electric shocks, by pressing a lever before other group members [15], with crowd competition for an exit in a fire. These experiments are arguably far removed from the environmental reality of a crowd crush. Other studies include 'waiting-room experiments' [16] such as an experiment on individual and group reactions to smoke seeping into a room from under a door [17]. The reactions of individuals acting alone, in going to and opening the door, were markedly faster than in a group of strangers, reflecting the social constraint on being the first to act.

Research of evacuations, such as Canadian office building evacuations conducted in the 1970s [18] and crowd ingress and egress research reviewed in a range of papers [19,20], has been directly concerned with the relationship between people, the time to escape and the safety of buildings in use, with particular reference latterly to stair safety [21]. This research [20] suggests that there has been a significant underestimation in international codes of times to evacuate high-rise buildings, such as offices, based on the main criterion used to predict escape times in codes: the width of exits and stairways. Research of evacuations has been useful in indicating design features such as tread and riser dimensions, handrail heights and dimensions [22] which contribute to safety and differences between the 'effective' or actual width used by evacuees and the objective width of the stairways. Some of these research findings have been incorporated into codes such as the US (NFPA) Life Safety Code. Other investigators [23] have worked on quantitative predictions of times to evacuate high-rise buildings. These predictions and the carrying capacity research have been predominantly concerned with the 'vertical' movement of people down stairways, phased evacuations of one or more floors such as the 'fire floor' in a predetermined sequence [24] or simultaneous evacuations of all floors. Predictions of escape times have been based primarily on offices and the predicted and monitored time to move down stairways, rather than through research and predictions based on the pattern of 'horizontal' movement on the upper floors prior to reaching a stairway, as well.

The central focus of much of the computer simulation research has been smoke and flame spread and/or assumed patterns of escape behaviour in relation to floor plans and escape routes. The aim is ultimately to use these computer models in risk or life safety assessments of plans at the design stage of buildings, either by computer literate experts or as a 'user friendly' microcomputer design aid. Much of this work has been sponsored by the US National Bureau of Standards (NBS). Examples of computer models are BFIRES-II [25], EVACNET+ [26], ASET [27], EXITT and EXIT89 [28], and there has been much recent discussion of these models [28,29]. One of the models of fire spread and human escape behaviour which has been influential in the UK has been that of 'escape potential' [30], in which the gradual reduction in escape routes available from different locations in a building over time is assumed to affect the predicted time taken to escape and the direction of people's movement. The computer programs, to date, appear to have been influenced more by the physical science (model A) than psychological model (model B), described earlier.

BFIRES represents the predicted pattern of human movement in terms of the information available to a person (the program being sensitive to parameters which include the floor plan configuration, occupants' initial spatial locations, mobility, familiarity with the layout and permissible levels of occupant density). In general, the models (e.g. RSET/ASET = Required Safe Egress Time/Available Safe Egress Time) concentrate on smoke and fire spread times, assuming the starting-point for movement to be the moment an alarm bell sounds. The initial human response time to an alarm is not part of the predicted evacuation time equation [31]. EXIT89 leaves the program user to specify an initial response delay time. Since the form of warning and delay time is not specified, in the rare evacuation tests conducted to validate these models [28], it is not clear what contribution this might have had on the time taken to evacuate. A recent information paper detailing a 'risk assessment model' being developed by the UK Fire Research Station (FRS) [32] states that it models 'fire development' and the 'behaviour of people', but provides no information on the latter. While these simulations provide considerable potential in risk assessment, their predictions concerning human behaviour need to be made explicit and require validation against existing and future research of human behaviour in fires and evacuations.

The fourth research strategy of field research (primarily using retrospective questionnaires and interviews with fire survivors) is more consistent with the psychological model of human behaviour (model B), described earlier. Much of this research has been sponsored by the NBS, US NFPA, FRS and, latterly, by the UK Home Office. This research has been conducted primarily by a handful of researchers over a period of some 20 years. Early research in this 20-year period in the UK sponsored by the FRS [33], and in the USA by the NBS [34], used a questionnaire methodology to examine the first, second and third actions engaged in and movement in smoke by people in samples of predominantly residential fires.

Research on behalf of the FRS statistically analysed sequences of actions derived from a sample of case studies of domestic, multiple occupancy and hospital fires, identifying three main stages of Recognition – Coping – Escape in people's reactions to fires which are often ambiguous in their early stages [35]. The action or 'behavioural sequence' statistical analysis method used was also subsequently adopted by others [36]. Case studies of major fires, such as the MGM fire in 1980 [37], have proved to have advantages over the earlier questionnaire surveys since it is far easier to relate the behaviour to the physical setting in which the fire occurred. A notable research study was conducted of the Beverly Hills Supper Club fire in the USA, in 1977 [38]; this study included an appraisal of people's behaviour, based on an extensive set of questionnaires and interviews with the fire survivors, the location of the fatalities and an appraisal of the 'escape potential' from different areas of the building [30].

The research during the 1970s has led to edited publications [39,40], and other sources of information on some of the research in the 1970s and early 1980s include a set of studies [41] and reviews [42,43], the latter

Exit choice behaviour

complementing other chapters in the *SFPE Handbook* [20,44,45]. A reprint of a collection of studies in 1990 [39] includes subsequent research on behalf of the FRS on psychological responses to visual information warning displays [46], and estimates by people of the speed of fire spread and a descriptive analysis of the likely pattern of movement of fire victims in the King's Cross fire [47], but excludes chapters in the first edition which relate most directly to physical movement and fire safety engineering design such as in Pauls chapter [18] and omits much of the relevant research conducted since the late 1970s.

The most recent human behaviour research sponsored by the FRS has extended the study of reactions to visual information warning displays [48] and addressed the evacuation of people with disabilities [49]. This is an area of increasing attention with the appearance of BS 5588: Part 8 [50] and the introduction to the NFPA Life Safety Code of a Fire Safety Evaluation Scheme (FSES) for residential boarding homes [51]. The FSES method of matching the evacuation capability of residents against the provision of staff assistance, means of escape and fire protection [52] has also been recommended for use in Sweden [53]. Research on 'assisted escape' [54] has also been conducted [55,56] and there are a number of reviews of life safety literature on people with disabilities [57,58,59].

5.5 EXIT CHOICE BEHAVIOUR

A review in 1977 of the research of human movement and behaviour in fires [60] argued that there had been insufficient research of behaviour in relation to the physical environment and recommended research on 'exit choice behaviour'. 'Exit choice' is the provision and distribution of more than one exit of sufficient exit width to allow a population of a specified maximum number (expressed as 'occupant capacity' in relation to floor space) time to reach safety in the event of one exit being obstructed by a fire (or smoke). The exit choice behaviour has to satisfy requirements concerning maximum travel distances for different types of premises via stages 1, 2, 3 and 4 of the escape route, as specified by fire codes and regulations [61], namely:

Stage 1: escape from the room or area of fire origin.
Stage 2: escape from the compartment of origin by the circulation route to a final exit, entry to a protected stair or to an adjoining compartment offering refuge.
Stage 3: escape from the floor of origin to the ground level.
Stage 4: final escape at ground level.

Research on exit choice behaviour has begun to appear. One field study experiment [62] found that the number of firemen evacuees running to each of five possible 'unfamiliar' escape routes from the sixth floor of a department store, in response to an 'alarm' (a rotating lamp and whistle), could be predicted most accurately by formulae based on the visibility of stair entrances, in contrast to the stair width or distance to exits. Research in a doctoral thesis on 'escape behaviour in fires' [63] and a three-year research programme (1985–8) on 'exit choice behaviour' on behalf of the UK Home Office [64,65] has essentially been concerned with reconciling research on movement (model A) and human behaviour (model B). Research completed in 1984 [63] included statistical analyses of sequences of behaviour in residential fires, the pattern of movement and travel distances moved by people in a hotel fire and escape behaviour in the Summerland leisure complex fire on the Isle of Man, in 1973, in which 3000 were present and 50 people died [16,66,67]. Factors found statistically in combination to influence the direction of people's escape behaviour in this fire were:

1. Role (staff or public).
2. Location (in relation to group members and exits).
3. Guidance to exits from staff.
4. Affiliative movement in or towards family groups.
5. Familiarity with escape route.

As in the Beverly Hills Supper Club fire [38], a major contribution to deaths was a delay of some 20 min from staff becoming first aware of a small fire, to people starting to move from the heavily populated Solarium in Summerland and the Cabaret Room in the Beverly Hills Supper Club following a dramatic escalation in the fire [14]. Staff members in the Marquee Showbar (MSB), Summerland, left by their familiar route to work, the fire exit and rear emergency stairway. The majority of the public left by their familiar route – the main entrance [16,66]. Deaths in the MSB were confined almost exclusively to groups whose members were together when first alerted (because of delayed response and the fact that groups move at the speed of the slowest member of the group). Parents in the MSB, separated from children (in the relatively safe Play Area), were most likely to survive because they started to move before the flow of other people from the upper floors [67], and this finding leads one to question the conclusion that parents looking for children added to the danger [68,69].

5.6 ESCAPE BEHAVIOUR FACTORS, ASSUMPTIONS AND PRINCIPLES

Research literature reviews [63], and one of five reports submitted to the Home Office [64,65], have suggested that the following factors should be

Escape behaviour factors, assumptions and principles

researched in studying the distance, timing and direction of (exit choice) escape behaviour:

1. Advice provided (existing guidance prior to fire).
2. Role in occupancy (e.g. staff or public).
3. Escape route familiarity and building layout.
4. Group dynamics and attachments.
5. Characteristics such as age, infirmity and disability.
6. Location and proximity to exit.
7. Information/communication on fire in progress.
8. Smoke obscuration (visibility, irritancy and toxicity).
9. Fire characteristics (such as heat and smell).
10. Exit signs.
11. Light levels and light sources.

The literature review in the initial stage of the Home Office research programme suggested that a number of common assumptions about escape behaviour should be examined. These assumptions, consistent with the physical science model A and historically with fire safety design regulations and standards, were as follows:

1. People's safety cannot be guaranteed since, in certain circumstances, they 'panic', leading to inappropriate escape behaviour.
2. Individuals start to move as soon as they hear an alarm.
3. The time taken for people to evacuate a floor is primarily dependent on the time it takes them physically to move to and through an exit.
4. Movement in fires is characterized by the aim of escaping.
5. People are most likely to move towards the exit to which they are nearest.
6. People move independently of one another (unless in a dense crowd).
7. Fire exit signs help to ensure people find a route to safety.
8. People are unlikely to use a smoke-filled escape route.
9. All the people present are equally capable of physically moving to an exit.

Listed below are an alternative set of escape behaviour principles which should guide fire safety engineering design. These principles were originally presented as research propositions derived from a review of the research literature [65], verified subsequently as principles consistent with a psychological model (model B) of escape behaviour. Each of the above nine assumptions corresponds to a contrasting principle, below:

1. Deaths in large-scale fires attributed to 'panic' are far more likely to have been caused by delays in people receiving information about a fire.
2. Fire alarm sirens cannot always be relied upon to prompt people to move immediately to safety.

3. The start-up time (i.e. people's reaction to an alarm) is just as important as the time it takes physically to reach an exit.
4. Much of the movement in the early stages of fires is characterized by investigation, not escape.
5. As long as an exit is not seriously obstructed, people have a tendency to move in a familiar direction, even if further away, rather than to use a conventional unfamiliar fire escape route.
6. Individuals often move towards and with group members and maintain proximity as far as possible with individuals to whom they have emotional ties.
7. Fire exit signs are not always noticed (or recalled) and may not overcome difficulties in orientation and wayfinding imposed on escapes by the architectural layout and design of an escape route.
8. People are often prepared, if necessary, to try to move through smoke.
9. People's ability to move towards exits may vary considerably (e.g. a young fit adult as opposed to a person who is elderly or who has a disability).

5.7 RESEARCH STUDY EXAMPLES

Most published accounts of people's behaviour in fires are at best descriptive, and at worst anecdotal. Most records of evacuations simply record the overall evacuation time of a building and few other details; such information is insufficient to form the basis for fire safety engineering design principles. The four examples of research which follow are based on detailed statistical and numerical studies documented primarily in extensive unpublished research reports; further details are provided in the references cited:

Example 1: Nurses' hall of residence fire.
Example 2: Department store fire.
Example 3: Lecture theatre evacuations.
Example 4: Underground station evacuations.

Examples 1–3 are taken from UK Home Office sponsored research [64,65,70]; and Example 4 was sponsored by the Tyne and Wear Passenger Transport Executive [71,72,73]. Examples 1 and 2 are based on mapping and statistical analyses of actions and movement in fires, coded from police witness statements. Statistical act sequence analyses and location sequence movement analyses conducted for these examples are not presented here. Due to the difficulties in recording the time to respond in fires, examples 1 and 2 concentrate on the direction of movement and, by inference, the timing of exit choice behaviour. Examples 3 and 4 are based on monitored evacuations with people's movement coded from video recordings; examples 3 and 4 cover the timing of behaviour more explicitly and, by inference, what would have occurred in a real fire. A series of related studies of action decision-making and exit choice behaviour, based on individuals participating in an interactive video disc simulation of a hall of residence on fire, are not considered here [64,65,74].

The central theme in the examples presented is the fact that T (time to escape in fires and evacuations) = $t1$ (time to start to move) + $t2$ ([time to move, indirectly or directly, to exits and along escape routes); i.e. $T = t1 + t2$ (model B integrated with model A). In contrast, current fire design standards, regulations and fire safety engineering design principles tend to assume escape movement is determined by $T = t2$ (i.e. model A or direct movement to and through exits).

Example 1: Nurses' hall of residence fire

Figure 5.2 illustrates where the fire began and the layout of rooms on the second floor of a nurses' hall of residence with a ground floor and five upper floors (each similar in layout); 51 occupants were present at the time of the fire, most of them asleep in bed in their individual rooms. The time of call to the Fire Brigade was 0057 hours. A statistical analysis of the sequences of actions (1084 actions) indicated the pattern of delayed response to ambiguous cues (e.g. shouts and/or the fire alarm), followed by investigation movement to open the room door. Those on floor 2 encountered thick smoke in the corridor. A statistical analysis of paths of movement (391 moves) coded in terms of 17 escape route locations (exits stairways, lift, floors, etc.) revealed three main paths of movement via the emergency (west) stairway, room window or main (east) stairway. Some backtracking behaviour occurred on the upper floors where people investigated and found their initial exit choice via the main stairway blocked by smoke.

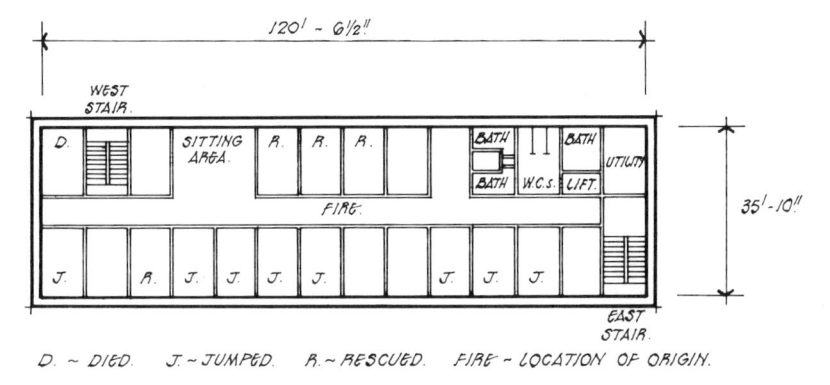

Figure 5.2 Plan of the second floor of the nurses' hall of residence.

Table 5.1 Floor level location and numbers of nurses escaping by different routes in the hall of residence

Floor level	Escaped via west stairway	Escaped via room window	Escaped via east stairway
5th	9	0	0
4th	5	3	0
3rd	11	0	0
2nd	0	12	0
1st	3	0	7
Total	28	15	7
	(56%)	(30%)	(14%)

Table 5.1 summarizes the three escape routes used by the 50 people who escaped. The numbers in each column of the table indicate that the emergency (west) stairway was used predominantly by those people starting on floors 3–5 above the fire floor, the window by those on the fire floor (2) and the main (east) stairway by those below the fire floor (1).

Of the 51 occupants, one died on floor 2 (D in Figure 5.2) (and 20 were injured); 28 left via the 'emergency' stairway (west); 15 left via room windows, of which 8 jumped from floor 2 (J in Figure 5.2); 7 were rescued by the Fire Brigade from windows (4 from floor 2: R in Figure 5.2); and 7 left via the main stairway (east) route. The conditions were undoubtedly worst on floor 2 where thick smoke was encountered, forcing people back into their rooms.

The length of the longest corridor on floors 1–5 was 34 m. The average distances moved by people outside their rooms were 15 m (those on the fire floor 2), 57 m (floors 3–5) and 63 m (floor 1). Analyses revealed no

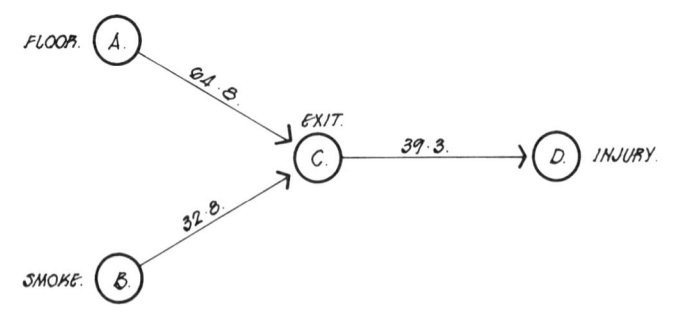

Figure 5.3 Hierarchical loglinear model (AC)(BC)(CD), fitting the data on the nurses' hall of residence fire with statistically significant associations indicated (LRC, $p < 0.001$).

statistical relationship between each individual's room location (i.e. travel distance to either exit onto the stairways) and the exit from floors people used (a finding supported in analyses of other fires in hostels, halls of residence and hotels examined [65]). In other words, people were not more likely to leave by a particular exit because they were nearer to it. Statistical relationships were found in a causal loglinear analysis [75], illustrated in Figure 5.3, between the (A) floor a person was on and (C) exit (escape) route used (Table 5.1); between the (B) smoke severity (density) encountered (as recalled) and (C) exit (escape) route used; and between (C) exit (escape) route used and (D) likelihood of injury.

Example 2: Department store fire

Figure 5.4 illustrates the proportions of the public and staff leaving by different routes in the Woolworth's department store fire in 1979. Figure 5.4 is a simplified version of a far more detailed statistical analysis of the moves and paths of escape movement between locations (1110 moves) by 132 individuals (63 members of the public and 69 staff), located primarily on floors 2 and 3. The movement was classified in terms of an index or 'location dictionary' of 27 locations, representing the exits, stairways, escalators, floor levels, etc. in the building. The location sequence (movement) analysis was complemented by an action sequence analysis, based on the 2463 actions of the same sample (132/500 = 25% of the estimated building population) [65]. Further details of the fire, which began among flammable polyurethane furniture in the furniture department adjacent to a restaurant, can be found in a report [76]. Ten people died in this fire (on floor 2). The fire was characterized by poor communications between staff, delayed call to the Fire Brigade, delayed alarm bell warning, delay in people in the restaurant starting to move, rapid escalation in the speed of fire spread and lack of sprinklers.

The main paths of escape movement, represented by arrows in Figure 5.4, were down stairway routes A or B; some people left via the escalator, a window or via the roof (with Fire Brigade assistance). The percentages at the bottom of the figure indicate the proportion of the public or staff leaving via the different routes. A notably higher proportion of the public left by stairway A (71%) than B (0%), in contrast to staff who were less likely to leave by A (27%) than B (41%). A number of the public also left by the escalator. Those on floor 3, above the fire floor, were exclusively staff. While approximately equal proportions of staff on floor 2 (63%) as public (71%) left by stairway A, A was blocked by smoke on floor 3. A door on route C just above floor 2 was locked. Table 5.2 indicates the difficulty the staff on floor 3 had in trying to leave by various routes: 34% left directly via B, and 38% tried first at least one other route before leaving via B;

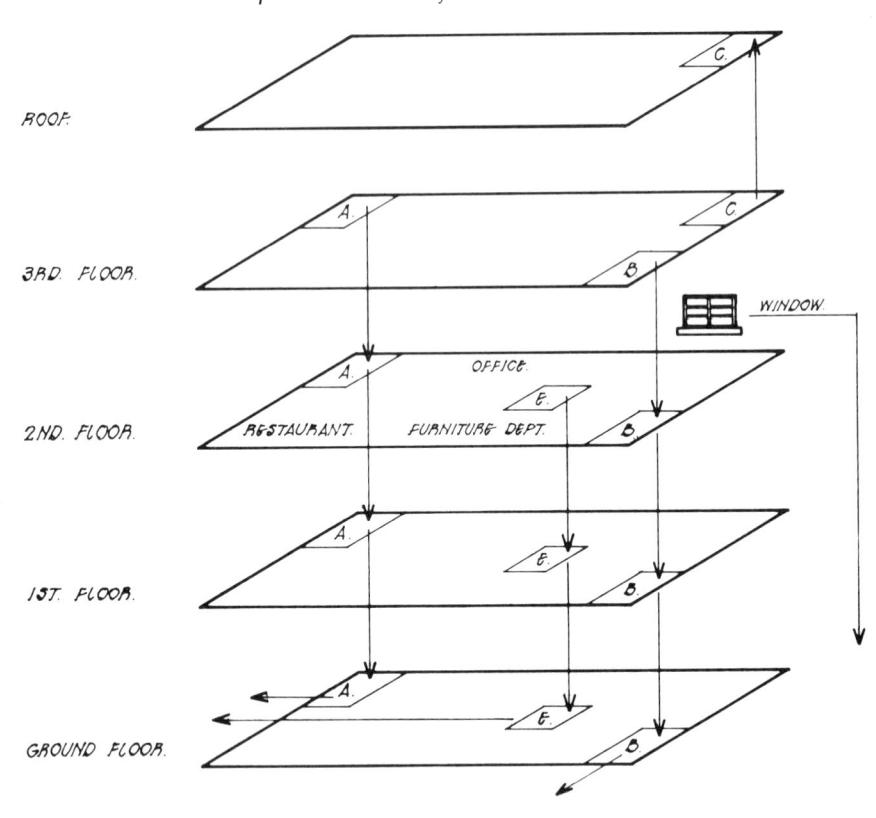

Figure 5.4 Proportion of public and staff using different escape routes from the second and third floors during the department store fire.

Key:
A = staircase A
B = staircase B

C = staircase C
E = escalator

Escape routes used:
Public = 71% staircase A
 22% escalator
 7% window

Staff = 27% staircase A
 5% escalator
 14% window
 41% staircase B
 13% roof

some staff tried as many as three exit routes before making their way up to the roof.

Figure 5.5 indicates that a statistical relationship was found between two main factors (A) Role (public or staff), (B) Floor (of location) and (C) the exit (escape) routes people escaped by in the fire. As the fire in

Table 5.2 Pattern of escape movement of staff on the third floor during department store fire

Escape pattern	Numbers	%
Stair B	11	34.4
Stair B → stair B	1	3.1
Stair A → stair B	3	9.4
Stair C → stair B	4	12.5
Goods lift → stair B	1	3.1
Stair A → stair C → stair B	2	6.3
Stair C → stair A → stair B	1	3.1
Stair C → window	1	3.1
Stair B → stair A → window	1	3.1
Stair C → roof	1	3.1
Stair C → stair C → roof	1	3.1
Stair B → stair A → stair C → roof	1	3.1
Stair B → stair C → roof	1	3.1
Stair A → stair C → stair C → roof	3	9.4
Total	32	

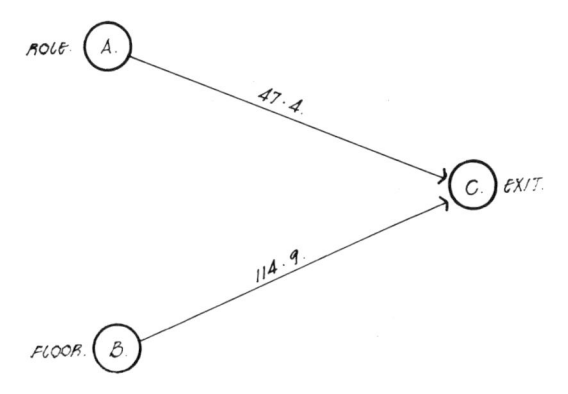

Figure 5.5 Hierarchical loglinear model (BC)(AC), fitting the data on the department store fire, with statistically significant associations indicated.

Note:
The figures next to the arrows are likelihood ratio chi-square changes (AC) $p < 0.001$, (BC) $p < 0.001$.

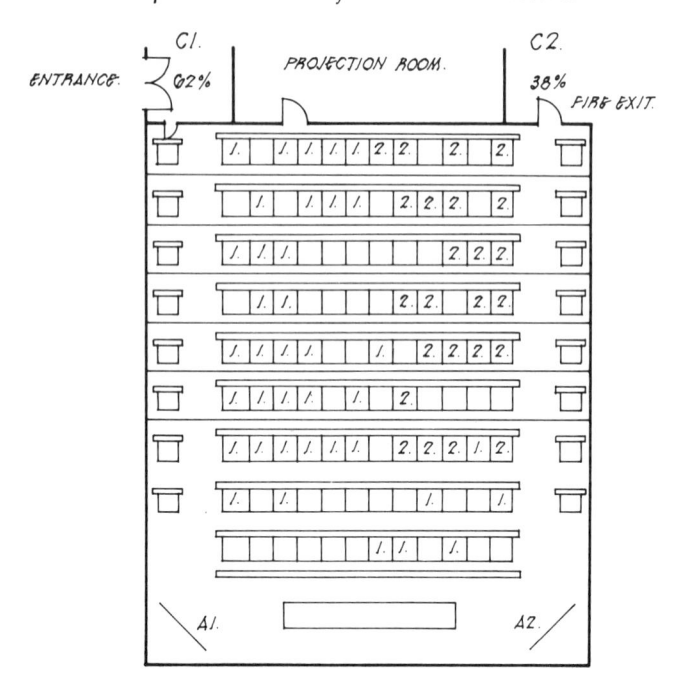

Figure 5.6 Front (F) lecture theatre seat locations of people using each exit (study 2), N = 63: C1 = used entrance; C2 = used fire exit.

example 1, above, did not have large numbers of staff in contrast to members of the public, the role of the person did not have a significant bearing on the route used. As in example 1, the floor people were on when first alerted (and in this case, by inference, the smoke and fire) influenced the route eventually used. The restaurant on floor 2 could be reached directly from the street via stairway A, suggesting a familiar frequently used 'frontstage' ingress and/or egress route for the public, unlike the 'backstage' staff areas and emergency stairways B and C. The analogy between audience and actors, public and staff, front and backstage of a theatre and a department store is a persuasive one. As staff are more likely to be in, have greater accessibility to and knowledge than the public of the 'backstage' region of a building [77], this is likely to have influenced the direction of escape behaviour by the two role groups.

Example 3: Lecture theatre evacuations

While the delay in people being aware of the fires in examples 1 and 2, above, undoubtedly influenced the timing ($T = t1 + t2$) of escape behaviour, it was not possible to measure this directly. Timing (T) was one of

Research study examples 73

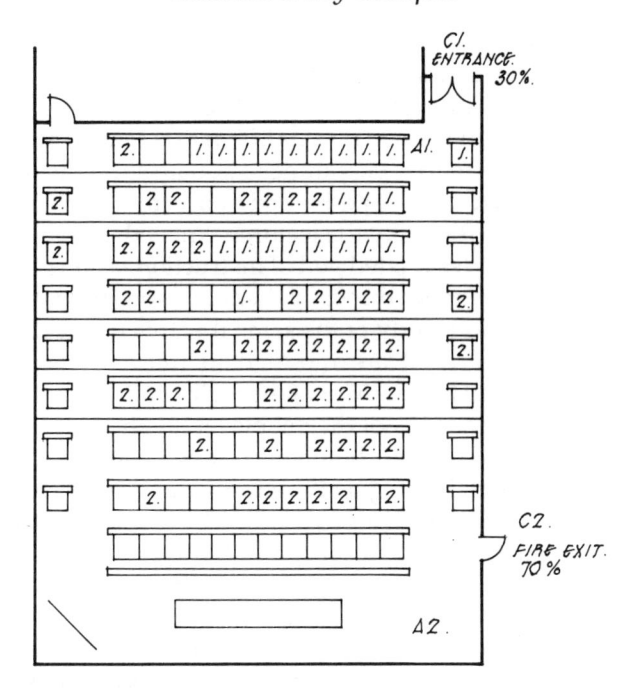

Figure 5.7 Rear (R) lecture theatre seat locations of people using each exit (study 2), N = 74: C1 = used entrance; C2 = used fire exit.

the main measures in the monitored evacuations represented by examples 3 and 4. Figures 5.6 and 5.7 illustrate the two lecture theatres which were the setting for example 3: a series of simultaneous evacuations of each theatre. The Front theatre (F) (Figure 5.6) and Rear theatre (R) (Figure 5.7) are on the ground floor of the same building and almost identical in design, except for the exit positions.

Assuming one exit is discounted, the Home Office Guide for theatres [69] paragraph 5.25 and Table D, p. 45) specifies that 'the evacuation time and the exit capacity of any exit or exit route should be determined' on the basis of a discharge rate of '40 persons per minute per unit width' (pum) of 525 mm in keeping with a standardized 2.5 mins as the calculated maximum permitted evacuation time, with this specified time and exit capacity in terms of the width and number of persons varying according to the class of construction: A = 3 mins, 120 pum, B = 2.5 mins and 100 pum, C = 2 mins, 80 pum.

In this example the width of each theatre is 9 m and the length is 10.5 m. In F (Front theatre) there is a 800 mm (32 in) wide entrance (single door) and 760 mm (30 in) wide fire exit at the back (both next to the building's front door). In R (Rear theatre) the 1300 mm (51 in) wide entrance (with double doors) at the back leads into the building and the

74   *Escape behaviour in fires and evacuations*

fire exit (again, 760 mm wide) at the front, leads directly to the outside of the building. The exits and 124 seats in each theatre are indicated in Figures 5.6 and 5.7. The numbers (1 or 2) in the squares indicate the location of the 63 occupied seats in F and 74 in R, when an alarm bell was sounded. In each theatre first-year undergraduate students (towards the end of the first term) were being addressed by a lecturer (15 min before the end of the 1-hour lecture). Only the lecturers knew the bell would sound and gave the same predesignated verbal instruction, asking the students to leave immediately (but not directing them to a specific exit).

A previous evacuation study of the same theatres a year earlier (study 1) and this study (study 2) are reported in full elsewhere [65]. The summary here is limited to a number of the main findings. Figures 5.6 and 5.7 indicate that whereas 62% of the people left by the entrance (C1) and 38% by the fire exit (C2) in F, the pattern of exit choice behaviour was reversed in R, with 30% leaving by the entrance and 70% by the fire exit.

In study 1, with a different sample, 55% left by the entrance and 45% by the fire exit in F and 0% by the entrance and 100% by the fire exit in R. The higher percentage of people leaving by the fire exit in R (study 1) is attributable to the fact that the lecturer walked across to the fire exit nearby in study 1 (R) and instructed everyone to leave that way. In F, study 1, the lecturer had not directed people to a particular exit.

Interviews with the lecturers and 21 students in F (study 2) and 21 in R (study 2) commenting after the event on their behaviour which when viewed on video revealed that the students regularly sat in the same positions in the theatres. Unlike those in R, those on the entrance side of F had generally not been aware there was a fire exit. In R the fire exit had been used regularly by these students (at least 50% of the time) to leave by after the weekly lecture. In F, the fire exit in F was never used this way.

A causal loglinear analysis [75] was conducted on the relationship between A: Seating Area/Location (A1 or A2 in Figures 5.6 and 5.7) × B: Distance Moved × C: Exit Used × D: Evacuation Time (Figures 5.8 and 5.9). Strong statistical associations were found between A × C, but not necessarily between C × D, as is reflected in Figures 5.8 and 5.9 (see the continuous arrows linking A × C). In F (study 2) the students had filtered out of the theatre at the same speed, up the inclining aisles on each side, with those at the front of F reaching the exits last (hence the continuous arrow B × D in Figure 5.8). In R (study 2) those near the entrance also left quickly, but so did the first people using the fire exit. However, some of those reached the fire exit in R from the entrance end, before those relatively near the fire exit (hence the continuous arrow B × C in Figure 5.9), by moving down the left aisle, rather than the more crowded right aisle. In F (study 1) a strong association was found between the exit used and the evacuation time (i.e. C × D) due to the fact that those sitting in the corner of F nearest to the entrance left first.

Research study examples 75

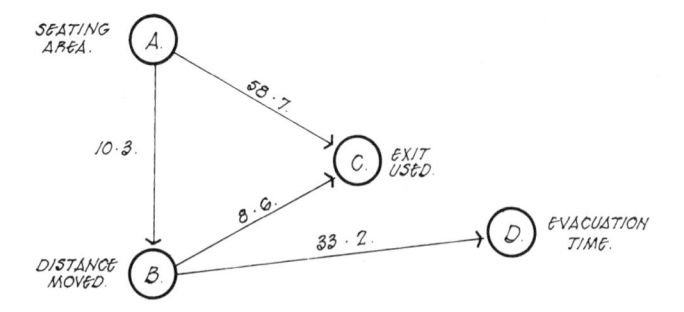

```
         SEATING
          AREA   (A)
                  |\
                  | \  58.7
             10.3 |  \
                  |   \
                  |    >(C) EXIT
                  |   / ⁸      USED
                  |  / 8.6
                  | /            33.2
         DISTANCE(B)---------------->(D) EVACUATION
          MOVED                                TIME
```

Figure 5.8 Best fitting loglinear model (AC)(BD)(AB)(BC), for front lecture theatre (study 2), with significant associations indicated.

Note:
The figures next to the arrows are likelihood ratio chi-square changes: (AC) p < 0.001; (BD) p < 0.001; (AB) p < 0.001; (BC) p < 0.01.

In summary, factors influencing the direction of movement were:

1. (Regular) seat position and proximity to exits.
2. Familiarity with exit routes (through regular use).
3. Seat and aisle layout in relation to exits.
4. Proximity of exits to the outside and orientation of the building.
5. Visibility of exits.
6. Instructions from the lecturer (directive or non-directive).

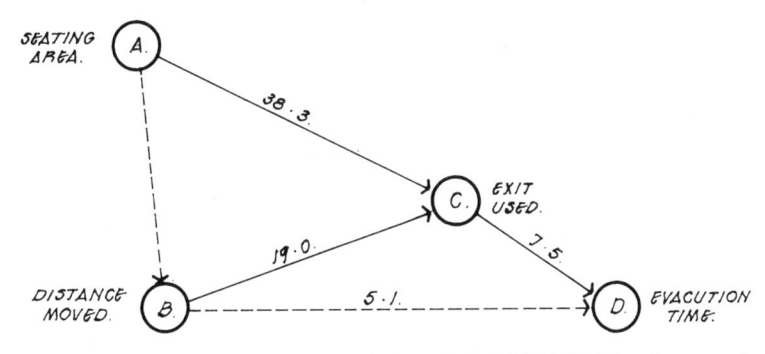

Figure 5.9 Best fitting loglinear model (AC)(BC)(CD)(ABD), for rear lecture theatre (study 2), with significant associations indicated.

Note:
The figures next to the arrows are likelihood ratio chi-square changes: (AC) p < 0.0001; (BC) p < 0.0001; (CD) p < 0.01; (ABD) p < 0.0. The interaction (ABD) is not shown: LRC = 5.2.

Table 5.3 summarizes for each theatre, F and R (study 1 and 2), the $t2$ range of D evacuation times (min: s) for the first and last person to leave by each exit, the $t2$ time span and the numbers leaving by each exit, C1 or C2. The larger overall $t2$ for F and R in study 1 (approx. 3 min), in contrast to half the time in study 2 (approx. 1:30 min), reflects the fact that the evacuation times were not determined exclusively or in a consistent fashion by the numbers of people, travel distances and exit widths. Although the $t1$ (time to start to move) was not recorded in study 1, the $t2$ times for study 1 in Table 5.3 suggest that people started to move later and took longer to leave (i.e. slower flow rate). The first person left in F (study 1) by the fire exit at 1.35 and in F (study 2) by the same fire exit at 0.21 s. The flow rates were all less than 40 persons per unit width per minute (pum): study 1: F entrance (ent.), 9.7 pum, fire exit (f.ex.), 12.3 pum; study 2: F ent., 22.0 pum, f.ex. 18.7, R ent., 18.1 pum, f.ex., 30.7 pum.

It is clear that the evacuation times were influenced at least as much by the response to the alarm, social behaviour and communications as the physical dimensions and design of the setting, although the two are interrelated. An important finding was that on average (ave) two-thirds of the evacuation time for individuals in both theatres, from the alarm sounding to exiting, was taken up by $t1$ (the time seated and in a standing position before moving to the exits); $t2$ (the time to move to and through the exits) constituted one-third of the average evacuation time. The ave $t1$ = approx. 30 s and ave $t2$ = approx. 15 s (theatre F ave $t1 = 31.7$ s, ave $t2 = 13.9$ s, ave T = 45.6 s; theatre R ave $t1 = 33.2$ s, ave $t2 = 15.8$ s, ave T = 48.9 s. These are averages, not overall T = $t1 + t2$ times, which were evidently much faster for study 2 than 1.

Table 5.3 Ranges of times to leave by C1 Entrance (Ent.) and C2 Fire Exit in front and rear lecture theatres (in minutes and seconds)

	Ent. first person	Ent. last person	Time span	N	Fire exit first person	Fire exit last person	Time span	N
Front theatre, Study 1	0:47	2:54	2:07	31	1:35	3:01	1:26	25
Rear theatre, Study 1		Entrance not used			0:50 approx.	3:00 approx.	2:10 approx.	77
Front theatre, Study 2	0:17	1:28	1:11	39	0:21	1:15	0:54	24
Rear theatre, Study 2	0:27	0:57	0:30	22	0:14	1:26	1:12	52

Example 4: Underground station evacuations

The final example further illustrates the significance of measuring and predicting evacuation times in terms of $T = t1 + t2$, rather than $T = t2$ by reference to the pattern of behaviour in five monitored evacuations of an underground station. The research conducted in 1990 on behalf of the Tyne and Wear Passenger Transport Executive, in association with the Tyne and Wear Metropolitan Fire Brigade, is recorded in detail elsewhere [71,72,73]. The study was modelled on the King's Cross fire in 1987. Figure 5.10 illustrates the layout of Monument Station, in Newcastle, which like the King's Cross underground station has two main escalators leading up to a central concourse level. A particular feature of the King's Cross fire were the poor communications with passengers. No alarm bell was sounded or warning given over a public address system; the station control room was not in operation. Transport police co-ordinated the evacuation of passengers from the lower levels of the station up the Victoria escalator to the concourse, where 31 people died, engulfed by the fire which spread suddenly from the other Piccadilly escalator [78].

In each of the Monument Station evacuations, the N/S escalators were sealed off by two firemen moving to the top and two to the bottom when the alarm sounded. This was equivalent to the Piccadilly 'fire escalator' in the King's Cross fire. Table 5.4 indicates the $t1$, time to start to move, for the people gathered in the concourse and at the bottom of the N/S fire escalator in each of 5 evacuations. Also given are the times to clear the station and a summary of the 'appropriateness of the behaviour'. The behaviour was recorded by videos on the CCTV system. The passengers on each occasion (like the students in example 3) were not informed beforehand that there was to be an evacuation exercise. The alarm bell was first sounded in each evacuation (evacuations 1–5). The type of evacuation warning was varied (left column, Table 5.4): evac. 1: alarm bell only; evac. 2: alarm bell + two staff on station; evac. 3: alarm bell + a twice-repeated 'nondirective' public announcement every 30 s (equivalent to a prerecorded message: 'Please evacuate the station'); evac. 4: alarm bell + two staff + 'directive' public announcements from a control centre (CC) operator using CCTV cameras co-ordinating the passenger evacuation via the concourse exits or from the lower levels onto trains; and evac. 5: alarm bell + no staff + more comprehensive directive public announcements about the 'suspected' fire, its location and actions.

Table 5.4 indicates that by varying the warning, the $t1$ (time to start), and therefore the T (evacuation time), varied dramatically. Whereas the alarm bell in evac. 1 had little effect on the passengers who either waited for more information or continued as normal (an extended $t1$), the

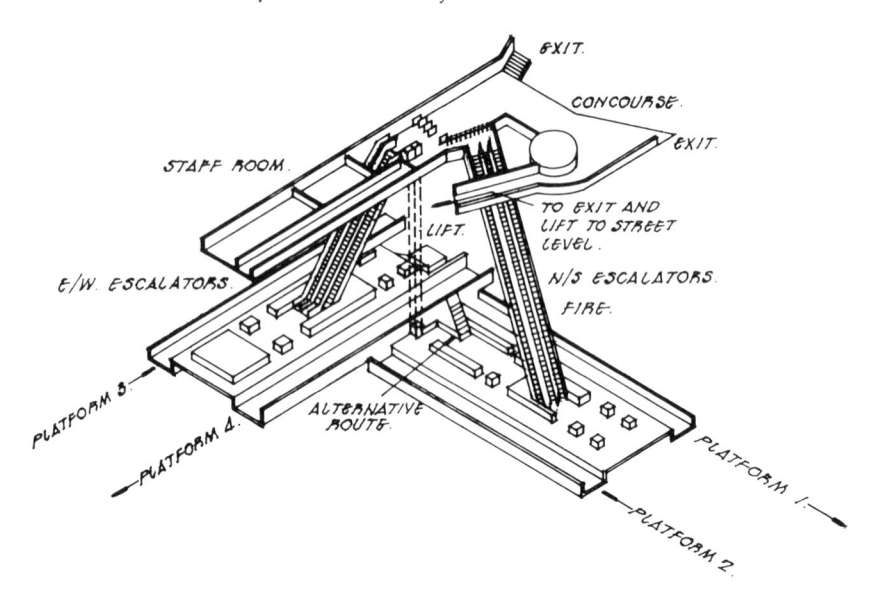

Figure 5.10 Monument Underground Station.

directive public announcements in evacuations 4 and 5 were extremely effective in reducing $t1$. (The number in brackets in evac. 5 represents two groups of latecomers with pushchair and pram.) The repeated public announcement in evac. 3 was ineffective at the bottom of the escalator. The member of staff at the platform levels in evac. 2 sent all 61 passengers up the E/W escalator (instead of to the trains) in exactly the same way as passengers had been directed up to the danger area in the King's Cross fire. The warning provision in evac. 2 (two staff and local access to a public address system) is consistent with the Fire Precautions (Subsurface Railway Stations) Regulations 1989, which, although produced after the King's Cross fire, curiously do not require a CC + CCTV + public address system (as in evacuations 4 and 5). The primary aim of the latter is to reduce $t1$ and send people in the safest direction. An effective CC system allows control operators a geographic overview of the station, which is virtually impossible for two staff 'on the ground' trying to cover a complex setting. The study suggests that evacuation times in the same physical setting can be reduced by at least one-half or even two-thirds (in this case, from approx. 15 min to 7 min or, perhaps, 5 min) by reducing the $t1$ component of $T = t1 + t2$.

5.8 IMPLICATIONS OF RESEARCH FOR FIRE SAFETY ENGINEERING DESIGN

In Chapter 6, five key characteristics of occupants crucial to escape design – i.e. sleeping risk, numbers, mobility, familiarity and response to fire

Implications of research

Table 5.4 Times and movement for the five evacuations at Monument Underground station (in minutes and seconds)

Evacuation	Time to start to move		Time to clear station	Appropriateness of behaviour
	Concourse	Bottom escalator		
1 Bell only	8:15	9:00	Exercise ended 14:47	Delayed or no evacuation
2 Staff	2:15	3:00	8:00	Users directed to concourse
3 PA system	1:15	7:40	10:30	Users stood at bottom of escalator
4 Staff + PA	1:15	1:30	6:45	Users evacuated
5 PA++	1:30	1:00	5:45 (10:15)	Users evacuated by trains and exits

alarm – are outlined and reference is made to the three escape strategies of egress, refuge and rescue. In the context of the numbers of people in different parts of a building and their mobility, escape route design dimensions of maximum travel distances and minimum exit widths are important. In keeping with other UK and international guides and codes, the Home Office [69] states that: 'the maximum evacuation time is an arbitrary figure used to determine distance of travel, width of escape routes and the number which theoretically would allow a given number of persons to escape to safety in a specified time.' The standard discharge rate of 40 pum used in calculations might be questioned [20], and a fire safety engineering risk assessment might wish to relate a design more directly to predicted walking speeds for a given population (e.g. a setting with many elderly people) or different crowd densities [79]. This chapter has been more concerned with communications and escape because of the effect that delays in warning the public have had in fire disasters [7].

In this respect, a broad definition of fire safety design has been adopted which, in relation to smoke and fire spread and fire protection systems covered in other chapters, should address the relationship between (A) Design × (B) Information Technology × (C) Management × (D) Buildings in Use. In this sense, principles 1–9 (section 5.6) are first-order (D) × (C) × (B) factors which should precede the second-order architectural, engineering and interior (A) Design detailing of a particular setting. The (A) layout of a setting and (B) warning system and messages need to be such that they can (C) be efficiently managed and (D) easy for the public to comprehend and use. All routes into and out of a building should ideally be part of the normal circulation routes. If this is not possible and issues of safety vs security often have to be reconciled, the lack of familiarity has to be redressed through appropriate warning systems, evacuation planning and practice.

The major problem remains, however, that of reconciling the physical science and psychological models. The closer integration of the two models (A and B) in a variety of areas is needed. An example is recent research which suggests the parity or equivalence between, and possible advantages of emergency 'wayfinding lighting systems', of low-level miniature tungsten filament or electroluminescent lamps or photoluminescent markings, over conventional high-level emergency lighting [80,81,82]. The important point to remember here, as in other areas, is that lighting should be considered not only in terms of movement, but building comprehension (i.e. as part of an information system).

At present the impact of the research presented on fire safety design codes has primarily been in the form of 'commentary', rather than substantive design 'recommendations' of a numerical kind (e.g. 'people need clear and accurate information as early as possible to ensure that all are able to reach a place of safety' [69]). BS 5588: Part 10 (1991) has a section on 'Wayfinding and Spatial Orientation' and also suggests that in shopping complexes 'directive' public address messages should take precedence over alarm bells and non-directive, pre-recorded messages. These suggestions have been directly influenced by the evacuation study cited in example 4, above. Similarly, POE video monitoring of evacuations used in examples 3 and 4 is in keeping with the kind of in-house monitoring needed if an integrated approach to fire safety design and management is to be achieved (see BS 5588: Part 10: 40 Review and Testing of Fire Safety Manual). Unfortunately, other BS 5588 documents (5, 50) do not include the same emphasis on regular monitoring and updating of evacuation and communication procedures specified in a fire safety manual tailored to a setting. A guide to evacuation management (including monitoring procedures) is urgently needed.

Research (including examples 1–4, above) suggests that travel distance has less effect on escape behaviour in certain types of setting (e.g. a hall of residence or hotel) than others (e.g. a theatre). Example 1 and research of the position of fatalities in corridors [65] suggests the potential role of rooms in hotels and halls of residence might be considered as fire-protected refuges, and the need to consider the circumstances in which people should necessarily be encouraged through a warning system to move from a room to an escape stairway (relevant to people with mobility difficulties).

While the evacuation times in the theatres (example 3) appear to be reasonable in terms of a 2–3 min criterion, it should be noted that:

1. The theatres were not full.
2. Two exits were used.
3. The times between studies 1 and 2 were variable.
4. The response time $t1$ to an alarm might well be far longer in a different type of social and physical setting (e.g. audience attending a play, less accessible final exits).
5. The relatively fast response to the alarm in example 2 was not the case in the more complex underground station setting (example 4: evacuation 1).

Circumstances might be considered in public settings where it should be possible to relax travel distances within strictly defined limits as a trade-off, where there is an effective information warning system and/or familiar public escape routes.

The research cited raises many questions relevant to current fire safety and crowd engineering design issues which have not been resolved. For example, following the Bradford and Hillsborough disasters, in what respect might the requirement that upper-division football stadia have all-seater, in contrast to standing facilities, increase rather than reduce $t1$? As in other settings, escape time predictions for spectator stands are closer to a $T = t2$ movement than $T = t1 + t2$ escape behaviour model [83]. However, there is a growing recognition of the need for a closer interface between crowd safety engineering and psychology [84]. What should the wording of public address messages be? A $T = t1 + t2$ rather than $T = t2$ model is as applicable to passengers in vehicles in train shuttle wagons in the Channel Tunnel as it is to a range of other settings, with a consequent need for the provision of a clear, prompt and accurate warning system of directive public address and visual display warning messages. The predominant engineering model for spectator or passenger movement needs to be reconciled with psychological imperatives.

Recently there has been a move towards the inclusion of $t1$ estimates in risk assessment probability equations comparing predicted evacuation times in relation to (F) fire spread times (see comparable psychological and evacuation time/fire development time models [31,42,85]). In recommending a time and probability-based fire safety code the elements of H = evacuation time which need to be established: $H = D + B + E/F$, have been similarly outlined [86]. Here $B = t1$, $E = t2$ and D = elapsed time from ignition to discovering or detecting the existence of the fire (e.g. through smoke detectors linked to an alarm bell). In view of the fact that warning systems have such a significant effect on evacuation times, it is surprising how little attention has been directed to their inclusion in discussion and numerical appraisal of combinations of fire safety engineering design measures likely to reach minimum, sufficient and 'equivalent' standards of public safety in the event of a fire. Attention needs to be given to the marriage between evolving communications technology which reduces $t1$ most effectively (i.e. distributed rather than centralized building 'intelligence' based on informative warning systems and public address communications from an efficiently operated CC using CCTV). Current computer-modelled simulations of evacuations in complex settings such as oil rigs (EGRESS [87]), and public settings (VEGAS: Virtual Egress, Analysis and Simulation [88]), have considerable potential but need to be validated very carefully against research of: human behaviour (actions, cognition, social factors) + human movement + design parameters + hazard development.

The 1991 draft Australian Building Fire Safety Systems Code [89,90] proposes a risk assessment model consisting of various subsystems relating not only to fire and smoke spread and building parameters, but occupant performance. This includes classification of building use and occupants, occupant communication and response and occupant avoidance times. Here: T_{me} (evacuation time) = T_r (response time) + T_p (preparation time) + T_m (avoidance time) [86]. Assuming that the $T_{me} = T_r + T_p + T_m$ predictions eventually used are derived on the basis of sound research data, this fire code could set an important international precedent for fire safety engineering design codes. There are signs of comparable fire safety engineering approaches being initiated in Canada, through the National Research Council, and in the UK, through the British Standards Institution.

At present there are no predictions in fire design codes of evacuation times in relation to different warning systems and messages, yet research indicates that these times should be set out in codes. Evacuation models and risk assessments which do not explicitly include a $t1$ component (i.e. are based exclusively on escape model A of $T = t2$) are fundamentally flawed. Research suggests that the 'margin of safety' [31], in terms of reducing evacuation times, lies more in reducing $t1$ (the time to start to move) than $t2$ (the time to move). A reduction in $t1$ gives people more time to find, cover the travel distance to and pass through an exit. A substantive database on the relationship between different warning systems and escape times is urgently needed. Ignoring this relationship could continue to prove fatal.

ACKNOWLEDGEMENTS

The studies presented were conducted in collaboration with Dr Michiharu Kimura and Dr Guylene Proulx. Their contributions are gratefully acknowledged. The views expressed are those of the author and not necessarily those of sponsoring bodies, namely: the Home Office (examples 1, 2 and 3) and the Tyne and Wear Passenger Transport Executive (example 4).

REFERENCES

1. Stollard, P. and Abrahams, J. (1991) *Fire from First Principles*, E & F.N. Spon, London.
2. Butcher, G. and Parnell, A. (1983) *Designing for Fire Safety*, Wiley, Chichester.
3. Shields, J. and Silcock, G.W.H. (1987) *Buildings and Fire*, Longman, Harlow.
4. Home Office and Department of the Environment (1992) *National Core Curriculum in Fire Safety Studies by Design*, IFE, IAAS, IBC, RIBA, The Institution of Fire Engineers, Leicester.
5. British Standards Institution (1991) *Code of Practice for Fire Precautions in the Design of Buildings, BS 5588, Part 6: Code of Practice for Places of Assembly*, BSI, London.
6. British Standards Institution (1991) *Code of Practice for Fire Precautions in the Design of Buildings, BS 5588, Part 10: Code of Practice for Shopping Complexes*, BSI, London.
7. Sime, J. (1992) Crowd safety management and communications in disasters, in *Management of Safety, Health and Environment* (MoSHE), Postgraduate Course, Module 2: Evaluation and Control, Manual No. 2, TopTech Studies/ Delft University of Technology, Delft, chapter 9, pp. 9.1–9.17.
8. Zeisel, J. (1981) *Inquiry by Design*, Brooks/Cole, Monterey, Calif.
9. Passini, R. (1984) *Wayfinding in Architecture*, Van Nostrand Reinhold, New York.
10. Arthur, P. and Passini, R. (1992) *Wayfinding: People, Signs and Architecture*, McGraw-Hill Ryerson, New York.
11. Sime, J. (ed.), *Safety in the Built Environment*, E. & F.N. Spon, London.
12. Garling, T. and Golledge, R. (1989) Environmental perception and cognition, in *Advances in Environment, Behaviour and Design* (eds. Z. Moore and G.T. Moore), Plenum Press, New York, Vol. 2, pp. 203–36.
13. Sime, J. (1985) Designing for people or ball bearings? *Design Studies*, **6**(3), 163–8.
14. Sime, J. (1980) The concept of panic, in *Fires and Human Behaviour* (ed. D. Canter), Wiley, Chichester, chapter 5.
15. Guten, S. and Vernon, L.A. (1972) Likelihood of escape, likelihood of danger and panic behaviour. *Journal of Social Psychology*, 87, 29–36.
16. Sime, J. (1983) Affiliative behaviour during escape to building exits. *Journal of Environmental Psychology*, 3(1), 21–41.
17. Latane, B. and Darley, J.M. (1968) Group inhibition of bystander intervention in emergencies. *Journal of Personality and Social Psychology*, **10**(3), 215–21.
18. Pauls, J. (1980) Building evacuation: research methods and case studies, in *Fires and Human Behaviour* (ed. D. Canter), Wiley, Chichester, chapter 13.
19. Pauls, J. (1984) The movement of people in buildings and design solutions for means of egress. *Fire Technology*, **20**(1), 27–47.
20. Pauls, J. (1988) Movement of people, in *SFPE Handbook of Fire Protection Engineering* (ed. P.J. DiNenno), NFPA/Society of Fire Protection Engineers, Boston, Mass., section 1, chapter 15.
21. Pauls, J. (1991) Safety standards requirements, and litigation in relation to building use and safety, especially safety from falls involving stairs. *Safety*

Science (ed. L.H.J. Goosens), special issue: 'Building and Environment Use and Safety', 14(2), 125–65.
22. Pauls, J. (1987) Are functional handrails within our grasp?, in *Public Environments* (eds J. Harvey and D. Henning), EDRA 18: Ottawa, Canada, Environmental Design Research Association Conference, EDRA, Washington, DC.
23. Kendik, E. (1986) Methods of design for means of egress: towards a quantitative comparison of national code requirements, in *Fire Safety Science: Proceedings of the First International Symposium* (eds C.E. Grant and P.J. Pagni), Hemisphere, Washington, DC.
24. London District Surveyors Association (1986) *Phased Evacuation from Office Buildings*, Fire Safety Guide No. 3, LFCDA, LDSA, London.
25. Stahl, F. (1982) BFIRES-II: A behaviour based computer simulation of emergency egress during fires. *Fire Technology*, 18, 49–65.
26. Kisko, T.M., Frances, R.L. and Noble, C.R. (1985) *EVACNET+: User's Guide*, Department of Industrial and Systems Engineering, University of Florida, Gainesville, Fl.
27. Cooper, L. and Stroup, D. (1985) ASET – computer program for calculating available safe egress time. *Fire Safety Journal*, 9, 29–45.
28. Fahy, R. (1986) EXIT89: an evacuation model for high-rise buildings, in *Fire Safety Science: Proceedings of the First International Symposium* (eds C.E. Grant and P.J. Pagni), Washington DC.
29. Kostreva, M., Wiecek, M.M. and Getachew, T. (1986) Optimisation models in fire egress analysis for residential buildings, in *Fire Safety Science: Proceedings of the First International Symposium* (eds C.E. Grant and P.J. Pagni), IAFSS, Hemisphere, Washington, DC.
30. Berlin, G.N. (1980) A modelling procedure for analysing the effect of design on emergency potential, in *Second International Seminar on Human Behaviour in Fire Emergencies, Proceedings of Seminar* (eds B.M. Levin and R.L. Paulsen), NBSIR 80–2070, National Bureau of Standards, Washington, DC.
31. Sime, J. (1986) Perceived time available: the margin of safety in fires, in *Fire Safety Science: Proceedings of the First International Symposium* (eds C.E. Grant and P.J. Pagni), IAFSS, Hemisphere, Washington, DC.
32. Phillips, W. (1992) *The Development of a Fire Risk Assessment Model*, BRE Information Paper 8/92, Building Research Establishment, Borehamwood.
33. Wood, P. (1972) *The Behaviour of People in Fires*, Fire Research Note 953, Building Research Establishment, Borehamwood.
34. Bryan, J. (1977) *Smoke as a Determinant of Human Behaviour in Fire Situations (Project People)*, Centre for Fire Research, National Bureau of Standards, Grant No. 4–9027.
35. Canter, D., Breaux, J. and Sime, J. (1980) Domestic, multi-occupancy and hospital fires, in *Fires and Human Behaviour* (ed. D. Canter), Wiley, Chichester, chapter 8.
36. Keating, J. and Loftus, E. (1984) *Post-fire Interviews: Development and Field Validation of the Behavioural Sequence Interview Technique*, NBS-GCR-84-477, National Bureau of Standards, Washington, DC.
37. Bryan, J. (1983) *An Examination and Analysis of the Dynamics of the Human Behaviour in the MGM Grand Hotel Fire*, NFPA NoLS-5, National Fire Protection Association, Quincey, Mass.
38. Best, R. (1977) *Reconstruction of a Tragedy: The Beverly Hills Supper Club Fire*, NFPA No.LS-2, National Fire Protection Association, Quincey, Mass.
39. Canter, D. (ed.) (1980) *Fires and Human Behaviour*, Wiley, Chichester.

40. Levin, B.M. and Paulsen, R.L. (eds) (1980) *Second International Seminar on Human Behaviour in Fire Emergencies, Proceedings of Seminar*, NBSIR 80-2070, National Bureau of Standards, Washington, DC.
41. Canter, D. (1985) *Studies of Human Behaviour in Fire: Empirical Results and their Implications for Education and Design*, Building Research Establishment Report L61, BRE, Borehamwood.
42. Stahl, F., Crosson, J.J. and Margulis, S.T. (1982) *Time-based Capabilities of Occupants to Escape Fires in Public Buildings: A Review of Code Provisions and Technical Literature*, Report NBSIR 82-2480, US Department of Commerce, National Bureau of Standards, Washington, DC.
43. Bryan, J. (1988) Behavioural response to fire and smoke, in *SFPE Handbook of Fire Protection Engineering* (ed. P.J. DiNenno), NFPA/Society of Fire Protection Engineers, Boston, Mass., section 1, chapter 16.
44. Nelson, H. and MacLennan, H. (1988) Emergency Movement, in *SFPE Handbook of Fire Protection Engineering* (ed. P.J. DiNenno), NFPA/Society of Fire Protection Engineers, Boston, Mass., section 2, chapter 6.
45. Purser, D. (1988) Toxicity assessment of combustion products, in *SFPE Handbook of Fire Protection Engineering* (ed. P.J. DiNenno), NFPA/Society of Fire Protection Engineers, Boston, Mass., section 1, chapter 14.
46. Canter, D., Powell, J. and Booker, K. (1988) *Psychological Aspects of Informative Fire Warning Systems*, Building Research Establishment Report BR 127, BRE, Borehamwood.
47. Donald, I. and Canter, D. (1992) Intentionality and fatality during the King's Cross underground fire. *European Journal of Social Psychology*, 22, 203–18.
48. Geyer, T., Bellamy, L., Max-Lino, R., Harrison, P., Bahrami, Z. and Modha, B. (1988) An evaluation of the effectiveness of the components of informative warning systems, in *Safety in the Built Environment* (ed. J. Sime), E. & F.N. Spon, London.
49. Shields, J. (1991) *Fire and the Disabled*, Fire Research Centre, University of Ulster, Final Report to Fire Research Station, Borehamwood.
50. British Standards Institution (1988) *Code of Practice for Fire Precautions in the Design of Buildings, BS 5588, Part 8: Code of Practice for Means of Escape for Disabled People*, 1988, BSI, London.
51. Lathrop, J. (ed.) (1985) *Life Safety Handbook*, 3rd edn, National Fire Protection Association, Quincey, Mass.
52. Nelson, H.E., Levin, B.M., Shike, A.J., Groner, N.E., Paulsen, R.L., Alvord, D.M. and Thorne, S.D. (1983) *A Fire Safety Evaluation Scheme for Board and Care Homes*, NBSIR 83–2659, US Department of Commerce, National Bureau of Standards, Washington, DC.
53. Hallberg, G. (1988) Evacuation safety in dwellings for the elderly, in *Safety in the Built Environment* (ed. J. Sime), E. & F.N. Spon, London.
54. Cooke, G. (1991) *Assisted Means of Escape of Disabled People from Fire in Tall Buildings*, Building Research Establishment Information Paper 16/91, BRE, Borehamwood.
55. Gartshore, P. and Sime, J. (1987) Assisted escape – some guidelines for designers, building managers and the mobility impaired. *Design for Special Needs*, 42, January–April, 6–9.
56. Sime, J. and Gartshore, P. (1987) Evacuating a wheelchair user down a stairway: a case study of an 'assisted escape', in *Public Environments* (eds J. Harvey and D. Henning), EDRA 18: Ottawa, Canada, Environmental Design Research Association Conference, EDRA, Washington, DC.
57. Pauls, J. (1988) *Egress Procedures and Technologies for People with Disabilities:*

Literature Review, prepared for US Architectural Transportation Barriers and Compliance Board, Washington, DC.
58. Sime, J. (1991) Handicapped people or handicapping environments? *Building Journal of Hong Kong China*, November, 84–92.
59. Sime, J. (1991) Accidents and disasters: vulnerability in the built environment, *Safety Science* (ed. L.H.J. Goosens), special issue: 'Building Environment Use and Safety', **14**(2), 103–24.
60. Stahl, F. and Archea, J. (1977) *An Assessment of Technical Literature on Emergency Egress from Buildings*, Report NBSIR 77–1313, National Bureau of Standards, Washington, DC.
61. Department of the Environment and the Welsh Office (1992) *Approved Document B: Fire Safety*, HMSO, London.
62. Horuchi, S. (1980) An experimental study on exit choice behaviour of occupants in an evacuation under building fire, in *Second International Seminar on Human Behaviour in Fire Emergencies* (eds B.M. Levin and R.L. Paulsen), Proceedings of Seminar, NBSIR 80–2070, National Bureau of Standards, Washington, DC.
63. Sime, J. (1984) Escape behaviour in fires: 'panic' or affiliation? PhD thesis, Psychology Department, University of Surrey.
64. Sime, J. (1987) Research on escape behaviour in fires: new directions. *Fire Research News*, 9, Spring, 3–5.
65. Sime, J. (1992) *Human Behaviour in Fires: Summary Report*, Building Use and Safety Research Unit (BUSRU), School of Architecture, Portsmouth Polytechnic, Central Fire/Joint Scottish Central Brigades Advisory Council, Home Office FRDG, London, Research Report No. 45, 1992.
66. Sime, J. (1985) The outcome of escape behaviour in the Summerland fire: panic or affiliation? in *International Conference on Building Use and Safety Technology: Conference Proceedings*, National Institute of Building Sciences, Washington, DC.
67. Sime, J. (1985) Movement towards the familiar: person and place affiliation in a fire entrapment setting. *Environment and Behaviour*, **17**(6), pp. 697–724, 1985.
68. Isle of Man Government (1974) *Summerland Fire Commission Report*, Isle of Man Government, Douglas.
69. Home Office and Scottish Home and Health Department (1990) *Guide to Fire Precautions in Existing Places of Entertainment and Like Premises*, HMSO, London.
70. Sime, J. and Kimura, M. (1988) The timing of escape: exit choice behaviour in fire and evacuations, in *Safety in the Built Environment* (ed. J. Sime), E. & F.N. Spon, London.
71. Sime, J., Proulx, G. and Kimura, M. (1990) Evacuation Safety in the Subsurface stations of Tyne and Wear Metro: Case Study of Monument Station, Stage 2 of a User Safety Evaluation (USE) on behalf of the Tyne and Wear Passenger Transport Executive, Jonathan Sime Associates.
72. Proulx, G. and Sime, J. (1991) To prevent 'panic' in an underground emergency: why not tell people the truth? in *Fire Safety Science: Third International Symposium* (eds G. Cox and B. Longford), IAFSS, Elsevier, London.
73. Proulx, G. (1991) Variation de l'information livrée aux usages pendant l'évacuation d'urgence d'une station de métro et développement d'une modèle de stress. PhD thesis, Faculté de l'aménagement, University of Montreal, Canada.
74. Powell, J., Creed, C. and Sime, J. (1988) Escape from burning buildings: a video-disc simulation for use in research and training, in *Safety in the Built Environment* (ed. J. Sime), E. & F.N. Spon, London.

References

75. Fienberg, S. (1977) *The Analysis of Cross-classified Categorical Data*, MIT Press, Cambridge, Mass.
76. Home Office (1980) *Report of the Planning/Legislations Subcommittee on the Fire at Woolworth's Piccadilly, Manchester, 8 May 1979, Joint Fire Prevention Committee of the Central Fire Brigades Advisory Councils for England and Wales and for Scotland*, Home Office Fire Department, London.
77. Goffman, E. (1959) *The Presentation of Self in Everyday Life*, Anchor Books, New York.
78. Fennel, D. (1988) *Investigation into the King's Cross Underground Fire*, Department of Transport, HMSO, London.
79. Fruin, J. (1971) *Pedestrian Planning and Design*, Maudep Press, New York.
80. Webber, G. and Hallman, P. (1988) Movement under various escape route lighting conditions, in *Safety in the Built Environment* (ed. J. Sime), E. & F.N. Spon, London.
81. Webber, G. and Hallman, P. (1989) *Photoluminescent Markings for Escape Routes*, Building Research Establishment Information Paper 17/89, BRE, Borehamwood.
82. Aizlewood, C.E. and Webber, G. (1992) Emergency escape routelighting: a comparison of human performance with traditional lighting and wayfinding systems, National Lighting Conference, Building Research Establishment, Borehamwood.
83. Home Office and Scottish Office (1990) *Guide to Safety at Sports Grounds*, HMSO, London.
84. Institution of Structural Engineers (1991) *Appraisal of Sports Grounds*, ISE, London.
85. Marchant, E. (1980) Modelling fire safety and risk, in *Fires and Human Behaviour* (ed. D. Canter), Wiley, Chichester, chapter 16.
86. Ramachandran, G. (1990) Probability based fire safety code. *Journal of Fire Protection*, **2**(3), 75–91.
87. Ketchel, N. (1992) Evaluating the Evacuability of Complex Structures: The Egress Code. Fourth Symposium on Oil and Gas Technology in a Wider Europe, Berlin.
88. Still, K. (1992) The lemming factor. *Health and Safety at Work*, December, 23.
89. Beck, V. (1991) Fire safety system design using risk assessment models: developments in Australia, in *Fire Safety Science: Third International Symposium* (eds G. Cox and B. Langford), IAFSS, Elsevier, London.
90. Beck, V., Clancy, P., Dowling, V.P. *et al.* (1991) Draft National Building Fire Safety Systems Regulation Review Task Force (C. Eaton), *Microeconomic Reform: Fire Regulation*, Appendix A, Building Regulation Review, Australia, May.

6
Fire escape in difficult circumstances

John Abrahams

6.1 INTRODUCTION

In the event of a fire the occupants of any building must be provided with a safe evacuation route which leads to a place of safety. This fire safety tactic of escape must be possible from every part of the building and available to all of the possible occupants. The crucial principle behind all means of escape is that it must be possible to escape between the time of discovery, or more correctly the time at which the occupants are made aware of fire, and the time at which conditions become untenable to the occupants due to the threat posed by the fire products – i.e. smoke, toxic gases and heat. Although this principle is clear, there is often confusion as to other underlying concepts, and until recently literature on means of escape has only been available in the form of codes of practice and guidance.

Many different codes of practice and guidance exist for particular building types; but unless there is an understanding of the basic principles of escape, there is a danger that architects, Fire Prevention Officers and Building Control Officers will apply them incorrectly. This is particularly relevant where the law might be framed in functional terms and the decision of the designers and those approving the design has to be seen to be sensible, effective and capable of withstanding cross-examination. Codes and guidance should be regarded as setting the standards of adequacy for a particular building type rather than as an absolute rulebook. The inclusion of means of escape within the new Approved Document B to the Building Regulations in England and Wales [1] gives designers the liberation to produce new or experimental designs, while the revision of the successive parts of BS 5588, 'Code of Practice for Fire Precautions in the Design of Buildings', provides the statutory authorities with a benchmark against which to assess the viability of such alternative proposals.

6.2 OCCUPANCY CHARACTERISTICS

If escape is to be effected between the time at which the occupants are made aware of fire and the time at which conditions become untenable to the occupants due to the threat posed by the fire products, it is essential to understand how fast the occupants will react and move. An understanding of the characteristics of the occupants will suggest their likely speed of travel, and in conjunction with the expected speed of fire spread, enable the architects to design adequate means of escape.

The nature and numbers of the occupants is probably more influential than certain of the physical design factors emphasized in escape codes and guidance. It is the interactions of the communication system with the occupants, the effectiveness of the signposting, the clarity of the internal layout and routes, the quality of fire safety training and response that will minimize the life risk from fire. Five key characteristics of the occupants can be identified, and these will now be considered:

1. Sleeping risk
2. Numbers
3. Mobility
4. Familiarity
5. Response to a fire alarm.

Buildings where people sleep are inherently more dangerous than those only used during the daytime. This is the single most important factor to recognize for the architect involved in the fire safety design of a building. When people are asleep, there is the opportunity for a fire to grow more before it is discovered, and even when discovered, the reactions of people who have been asleep will be much slower. Many fires are simply prevented from becoming much more than a minor incident by prompt preventive action or by simple extinction. Moving the clothes-horse further away from the heater when things smell hot or stamping on the small, glowing cinder that spits onto the carpet are familiar examples of normal domestic life. The combination of the lack of preventive actions and the slow response time to a fire make sleeping risk a particular feature to be considered. The Fairfield Old People's Home at Edwalton, Nottinghamshire, was severely damaged by fire in 1974 and 18 people died. The fire was thought to have been started by one of the residents smoking in their bedroom, while most of the other residents were asleep; it spread rapidly through the continuous ceiling void and was not detected until in an advanced stage. The greatest damage and loss of life ensued at the opposite end of the building from the start of the fire.

To plan an adequate means of escape the designer needs to know how many people will be in the building and where they are likely to be located. This will depend upon the building's function, but the architects must remember that a building designed for one purpose may well be used for another. For example, in one Hampshire school the fire service were horrified to find that the double height 'street' down the middle of the building was not just used for daytime circulation as they had expected, but was being used for such diverse evening activities as beer festivals. In the Republic of Ireland the most serious fire in recent years was on the night of St Valentine's Day 1981, when 48 young people died in a fire at the Stardust Disco in Dublin. One of the main problems was the sheer number of people in the disco at the time. Large numbers of people should not have been a problem, but combined with a total lack of any form of Building Regulations, inadequate local authority supervision and the owners' untrained advisers, they contributed to the disaster. The management failed to give the alarm and commence evacuation immediately the fire was noticed. Instead the occupants watched the unsuccessful attempts of two employees to extinguish the fire. Then without warning it became an inferno, the fire spreading exceptionally rapidly due to poor seating, wall lining materials and low ceilings.

It has already been stressed that people must be able to escape from the danger areas before they are overcome by the smoke and heat from the fire. However, people will escape at different speeds and there is no perfect figure which the designer can use. Some of the occupants may be disabled, encumbered or drunk. The worst cases will be those where assistance is essential if they are to move at all. Further complications can occur when considering orthopaedic patients will full traction equipment, or intensive-care patients on life support systems where it is necessary to carry out the evacuation and sustain the support systems. In between these two extremes lies the majority of the population with limited mobility. The designer will need to assess what proportion of occupants may be unable to move away readily from a fire and plan accordingly.

If the occupants are familiar with a building, they will find less difficulty in escaping from fire than those unfamiliar with their surroundings. In a strange building people will instinctively try to escape the way they came in, and it may be hard to persuade them to escape via 'official' designated escape routes if these are in the opposite direction. Therefore normal circulation and exit routes should always be regarded as escape routes. Escape routes which are not normally used, and only available in an emergency, should be avoided if at all possible. If they are unavoidable, then they will need explicit signposting.

Familiarity will vary with building type. In a normal domestic situation the occupants will be very familiar with the layout for their own house or flat. Similarly, office buildings and factory buildings will probably have a stable workforce that will be familiar with access and exit routes. Problems are likely to occur, however, in places like hotels and hostels

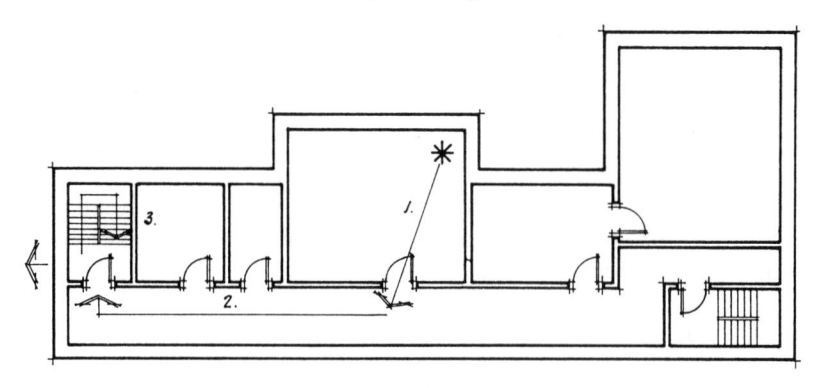

CELLULAR PLANNING.

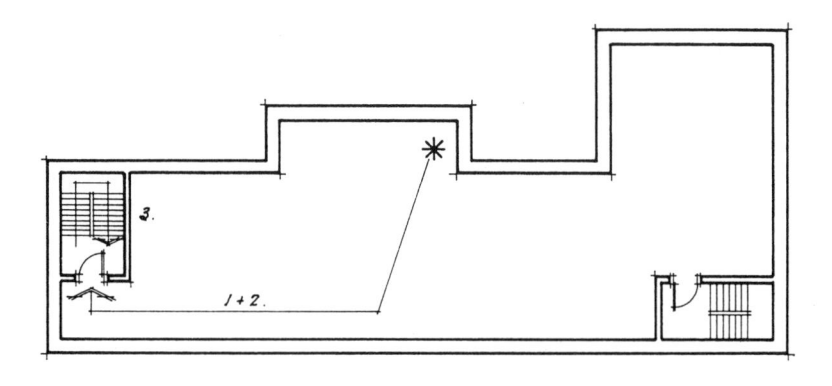

OPEN PLANNING.

Figure 6.1 Stages of escape.

where the residents may only stay a short time. The problem is particularly severe in clubs and cinemas where the occupants can be unaware of which level they are on, as well as the location of exit routes. The likely response to a fire or a fire alarm has to be considered as another feature. When a fire occurs or an alarm sounds, a variety of actions may take place. In a building where there is a well-disciplined staff with a planned evacuation strategy, the response will be markedly different from the building which contains people who may be unwilling or unable to appreciate the danger. In this context, an office building or an acute/surgical hospital may be safer examples, while a small home for the mentally handicapped, a conference centre or student residential building could offer more worrying examples.

6.3 ESCAPE STRATEGIES

The classical means of escape from a cellular building are considered in stages; these are:

Stage 1: escape from the room or area of fire origin.
Stage 2: escape from the compartment of origin by the circulation route to a final exit, entry to a protected stair or to an adjoining compartment offering refuge (Figure 6.1).
Stage 3: escape from the floor of origin to the ground level.
Stage 4: final escape at ground level.

However, it is not sufficient just to consider means of escape as a series of protected routes whereby people can travel from any point in the building to a place of safety by their own unaided efforts. Such a definition gives rise to a number of problems, for while many people can move adequately in evacuation, many disabled, chronically sick or sedated people will need assistance in evacuation. Therefore there are two basic escape strategies, of which the first, egress, is the simple direct escape from the building when the alarm is sounded.

The second strategy is that of refuge, where the structural fire containment of the building is used to provide a place of safety within the building, so that evacuation takes place from the compartment where the

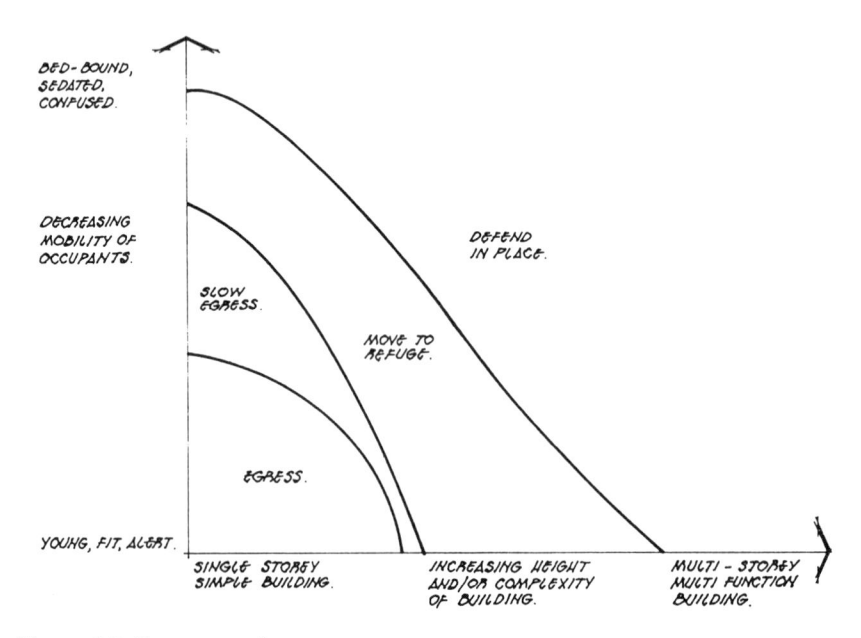

Figure 6.2 Egress vs refuge.

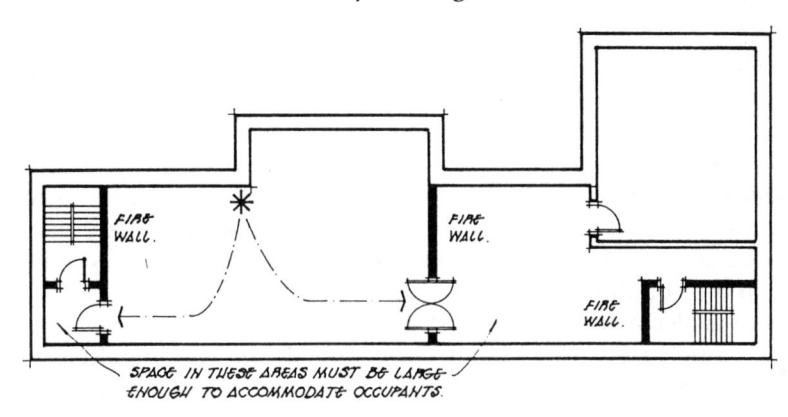

Figure 6.3 Horizontal escape to refuge.

fire started to an adjoining compartment (Figure 6.3). Clearly this is only acceptable when it is possible to continue further evacuation without returning through the compartment of origin. When the occupants are less mobile, disabled or incapacitated, the concept or strategy of 'refuge' is often more appropriate than that of 'egress'. This is particularly pertinent in larger and more complex buildings, where it is easier to design separate containment or sub-compartments as receiving compartments within a large building (Figure 6.2). It may also be necessary to design refuge areas adjoining vertical communication routes (lifts and stairs), so that it is possible to await rescue in safety. There are a number of codes of practice and design codes which now rely on or embody these principles; in particular, those codes for hospitals [2].

In the case of very limited mobility (e.g. operating theatre departments, coronary care units), a third strategy, that of defend in place, may have to be considered. This will require good fire containment (at least 1 h) and the compartment must also have its own separate, independent air supply and extract system and a protected electrical supply.

An extreme example of escape where a 'first principles approach' was essential occurred in a diving diseases treatment centre. This medical facility acts as an accident centre for divers suffering from 'the bends', and for cases of monoxide poisoning. It also offers treatment for a variety of conditions which benefit from a compressed oxygen-rich atmosphere. In this particular case, egress or movement could easily result in death or severe illness, therefore the centre has a number of escape strategies which range from the simple evacuation strategy when no diving is taking place to the 'no-stop diving' fire strategy which relies on achieving fire safety through the other tactics. A principal risk in the centre is the oxygen and oxygen enrichment and there have to be tight fire prevention controls in the chamber areas.

It is interesting to note that when these strategies were being developed increased reliance was placed on the other fire safety tactics of containment, extinguishment, communications and, above all, prevention. From the regulating authorities' viewpoint, such ideas based on the concept of equivalency or 'trade-offs' are not always easily accepted. At present the Building Regulations and Fire Precautions Act 1971 place an emphasis on structural fire resistance and physical fire safety matters, with perhaps insufficient consideration given to the fire safety management of the building or establishment.

There is also a fourth escape strategy which can be adopted as a last resort, namely rescue, by persons from outside the building. Rescue can be considered in the case of small buildings, but it is neither reliable nor commendable. The evacuation of the occupants by ladder may have to be considered if the building has only one stairway, but this is patently not suitable for other than small numbers of people and low-rise buildings. Also because it is hard to rescue the disabled, the chronic sick and infirm, design to facilitate rescue must be regarded as a helpful feature rather than the principal escape strategy. Rescue as a possible means of escape should be regarded as a bonus measure rather than a planned strategy for the larger and more complex building. Its relevance is with single-staircase buildings, especially where in addition there is little or no internal fire division or 'refuge' provision. In houses, for rescue to be successful as a design element it must include a full package of detailed window design, balcony design and access for fire service ladders, together with a certain amount of provision in respect of making upstairs rooms capable of acting as a refuge, even if only for a limited amount of time.

6.4 'FIRST PRINCIPLES' APPROACH TO ESCAPE

In complex or unusual buildings the design team may need to take a first principles approach to the design of the means of escape. This may involve the appointment of a fire safety specialist who can consider escape as part of a series of different packages of fire safety precautions, each of which will provide an equivalent level of safety. Lack of resources in the statutory authorities dealing with means of escape, means that a 'first principles' approach, as opposed to a code of practice solution is unlikely to be offered to an applicant.

Certain types of buildings (e.g. laboratories, secure hospital units) do not have any code of practice for means of escape and for these buildings a sensible strategy of evacuation has therefore to be developed as part of the architect's brief. The means of escape can then be developed along with the plan. The idea of designing the means of escape for a prison or forensic psychiatric unit is guaranteed to raise a smile, the objective being to keep the occupants under supervision at all times. In two units within the South-West Regional Health Authority the building is not only effectively sub-compartmented to limit the numbers that might need to be evacuated, it has also been necesssary to provide safe external compounds as the 'place of safety' into which they can escape. The most difficult problem was in the design of adequate means of escape for the control base staff who have control over the interlocking doors and electric locks on the final exits. These people need to continue at their posts until the evacuation is complete. The interlocked doors have necessitated the installation of a short dry riser running from the outside to each of the staff bases in the ward sections.

Historic buildings are another group where a first principles approach may be adopted with good effect. The successful approach can range from providing pressurized staircase enclosures instead of fire lobbies on single-staircase buildings to the careful detailing of fire-resisting screens. In Bristol City Library, for example, a perfect art nouveau interior has been retained while, at the same time, being upgraded in terms of fire resistance, the expert opinion of the Timber Research and Development Association (TRADA) being used for an assessment of fire resistance.

Evacuation problems can also result from the varied nature of the project. In one converted warehouse the developer sought to provide an art gallery, a café, a cinema, bookshop, office accommodation and a caretaker's flat. It is essential that first principles are used when this level of complexity exists.

To fully understand the calculations that will be involved in a first principles approach it is necessary to return to our initial statement, that the means of escape will have to be effected between the time at which the occupants are made aware of the fire and the time at which the conditions become untenable. Therefore it is necessary to consider both those factors which will determine the spread of the fire and those occupancy characteristics which have already been mentioned. The likely speed of spread is related both to the design and the management of the building, and together these will determine the potential fuel and smoke load. The probable speed of fire spread should not be underestimated and smoke logging can occur within a very short time – it is this time which is critical to the occupants for escape. In a complex building it is essential to have the best possible type of fire communications system. It is often necessary not only to signal that there is a fire, but also its point of origin, the affected zone or fire compartment and the preferred route of escape for the occupants. Greater levels of information and control can then be built into such systems, such as the closing of fire doors, shutting door supply air, as well as alerting the fire-fighting/rescue teams and the local facilities management.

It is not surprising that a lot of the work of fire safety engineers has followed the development of contemporary fire alarm and detection systems in the use of microprocessor and computer systems. The systems are capable of analysing and dealing successfully with a large number of variables. If the movement characteristics of different groups of people capable of different movement velocities, and the capacity of buildings in terms of the escape routes, stairways and final exits, are added to the initial calculations of fire development and smoke propagation, it is possible to develop integrated models of evacuation.

It may be pertinent in the design of very large schemes or large spaces to seek the assistance of computer modelling techniques. In the development of the Stansted Airport terminal building, Ove Arup were commissioned to develop a fire-engineered solution. This included a portrayal of smoke spread from a 'standardized' fire using a computer model and imposing a random distribution of individuals with different movement capabilities. The modelling is able to demonstrate the feasibility of the last disabled person struggling through the exit before the black cloud engulfs the area. The modelling is a tool to enable the designer to establish the suitability of large-volume areas and escape, and therefore one aspect of a first principles approach.

Despite these levels of sophistication, it is essential that both architects and the statutory authorities have a good appreciation of the conceptual aspects of escape strategies. Strict and slavish obedience to a code of practice is fraught with danger, if only that of complacency. There are many schemes which never quite fit into the categories envisaged by the drafters; there are schemes which contain a multiplicity of uses; and there are schemes where travel distance is not a suitable measure for escape, and it is the time available for escape which is critical.

REFERENCES

1. Department of the Environment and Welsh Office (1992), *Approved Document B: Fire Safety*, HMSO, London.
2. Department of Health (1980–), *Firecode Series*, HMSO, London.

7
Principles of fire containment

H. L. Malhotra

7.1 INTRODUCTION

As explained in Chapter 2, protection against fire requires a number of measures to be taken and one of these is provision for fire containment. It is assumed that measures to prevent a fire have not been successful: a fire has started and it is in the process of spreading. The purpose of fire containment is to prevent the fire spreading beyond defined boundaries and to safeguard the other areas so separated. The areas to which the fire is confined may be termed as a fire cell, a fire zone or compartment. This may be a complete building; a complete floor of a building; a vertical enclosure; or a specified part of a building. The objective of fire containment is achieved by compartmentation, with the dual purpose of preventing a fire escaping from the compartment of origin and preventing it entering into other compartments. Some compartments do not present much risk of a fire (e.g. escape routes), yet they still need to be designed to prevent the entry of a fire. Fire containment has a number of objectives; some of these are:

1. Limiting the risk to occupants.
2. Protecting essential areas such as escape routes.
3. Cost-effective use of escape provisions.
4. Reducing the damage potential of a fire.
5. Facilitating the control of fire.
6. Reducing the possibility of conflagration.

Fire containment is generally achieved by constructional measures or other inbuilt provisions which have the effect of preventing the consequences of a fire spreading beyond defined boundaries. There are two types of fire effects to be considered: the generation of heat, and the production of fire gases such as smoke and toxic products (Figure 7.1). Passive fire containment implies that the constructional features are such that the adverse effects are kept within the fire zone and by maintaining

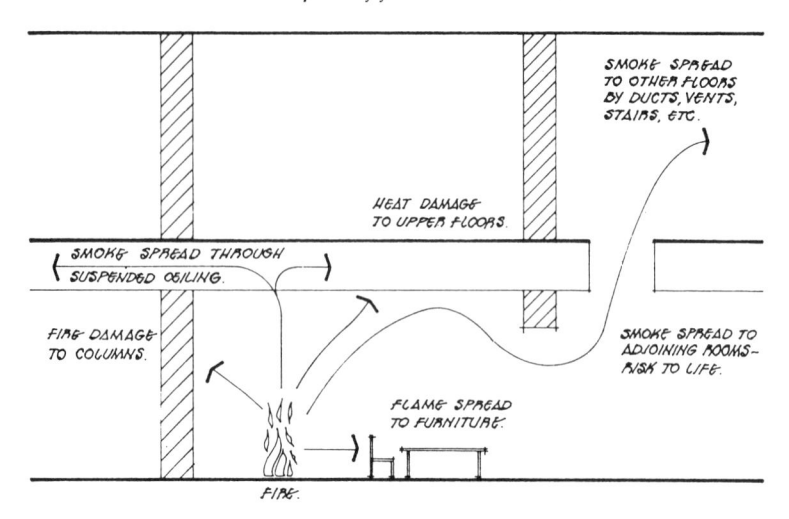

Figure 7.1 Threats from heat and smoke.

the integrity of the zone boundaries their spread is prevented (Figure 7.2). Two points worth noting in this context are, first, that the containment provisions have little effect on conditions within the fire zone; and secondly, that the zone boundaries may successfully contain a fire but would very likely need to be replaced afterwards. Fire containment is

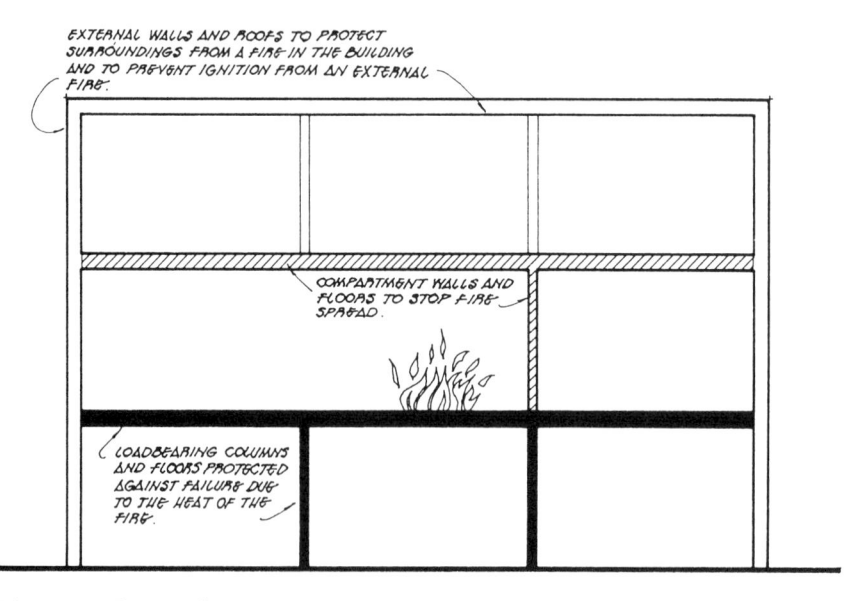

Figure 7.2 Passive fire containment.

Historical aspects

achieved by considering the natural division of spaces within a building and the possibility of treating them as separate compartments and designing the boundaries to withstand the effects of a fire of known severity. Fire containment is therefore a combination of compartmentation concepts with fire-resistant construction.

7.2 HISTORICAL ASPECTS

Historically fire containment has been the mainstay of fire protection through the centuries and was perhaps one of the main requirements in the early control systems. Separation of buildings or parts in different occupancies was an acknowledged goal of the early approach developed following the Great Fire of London in 1666. At that time property protection was considered to be more important than life safety and therefore the emphasis tended to be on measures which safeguarded properties from each other. The first measure following the Great Fire was to require houses to be separated from each other by brick walls at least 225 mm thick.

In the nineteenth century large warehouse-type buildings were constructed in the cities and a number of fires were experienced in them. Perhaps the most significant of these was the Tooley Street fire in London, in 1875, when the then fire chief, Eddie Shaw, is supposed to have said that his brigade could not have dealt with a bigger fire. The size of the warehouse was 250 000 ft^3 (+7000 m^3) and this is assumed to have been the basis for the upper limit for compartments in storage buildings.

The early history of the development of the prescription in the by-laws for the compartment size limits for different categories of buildings and the relationship with the height is not well recorded. The Fire Grading Report, published in 1946 [1], analysed the questions related to fire safety in buildings and studied the rules used in other countries, particularly North America. It took the limit of 250 000 ft^3 used by the London County Council as the upper limit and suggested a comprehensive classification system which related the occupancy type, the nature of the construction and facilities for fire control. It used seven types of construction ranging from 4 hour, non-combustible constructions to a combustible construction having no more than 30 min fire resistance. The compartment sizes varied from unlimited to 25 000 ft^2 (+2000 m^2), the unlimited size was accepted only for low fire load occupancies. For trade and storage buildings, the use of an automatic sprinkler system allowed the sizes to be doubled. Unfortunately, no reasoning was given for the compartment size subdivisions except that this reflected the US practice.

These recommendations were partially included in the postwar by-laws but became part of the comprehensive system devised by the Building Regulations in 1962–5 in mainland UK, with the exception of a general recognition of the contribution of a sprinkler system in reducing fire severity. The resulting system was very complex with the occupancy, building height and fire resistance related in an unsystematic manner. In a report to the DoE in 1987 on Fire Safety in Buildings a strong recommendation was made for simplifying the system [2].

7.3 FIRE COMPARTMENTATION NEEDS

It is necessary to consider the needs for compartmentation in order to understand the system as practised today and the changes which are taking place. The concept of fire containment can be taken to mean either the prevention of the spread of a fire, with its messengers of heat and gases, from the zone where it occurs to another separated part or, equally, the prevention of a fire entering a protected zone. This approach leads to the recognition of the following four objectives:

1. Subdivision of a building into zones/compartments in order to keep a fire to a manageable size.
2. Separation of different buildings, or parts of a building, to prevent a fire in one occupancy affecting another.
3. Physical separation of buildings to prevent conflagrations.
4. Keeping special areas safe from the effects of a fire.

The subdivision of a building on the basis of life safety would aim to limit the number of people at risk at any given time, allowing people in non-fire zones a longer time to escape, to provide a temporary refuge to utilize the concept of phased evacuation and therefore simpler design for means of escape. These concepts are particularly relevant in buildings with large occupant density, or where the occupants have special escape problems. Where a building is divided only into zones, the occupants in a single zone are immediately at risk at the start of a fire and need to utilize the means of escape (Figure 7.3). Others can stay temporarily in their place without danger from the effects of a fire. In health care buildings the occupants have difficulty in making their escape quickly and therefore require longer time to reach the outside of a building. This can be taken care of by providing the buildings with compartmentation at each floor level for rapid horizontal movement to a safe area, followed by a progressive movement to the outside of the building.

In most cases, life safety considerations require protection varying from a few minutes to perhaps a maximum of 60 min or so in large buildings. This means that if life safety is the only consideration, the maximum duration for effective compartmentation can be 60 min or less. However, if the safety of fire-fighters has also to be considered, and this needs to be recognized as a requirement in some cases, then some buildings may demand higher standards. The collapse of tall buildings is also a hazard to

Fire compartmentation needs 101

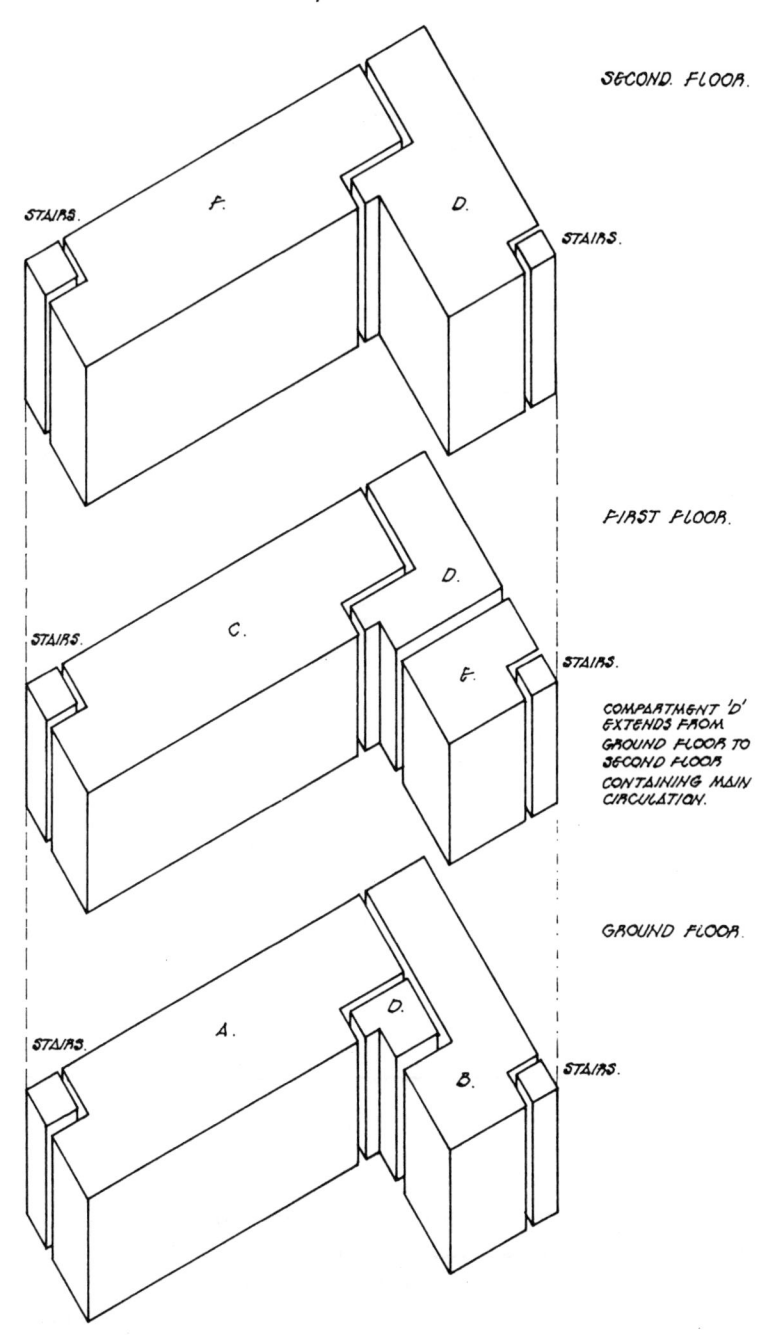

Figure 7.3 Compartmentation.

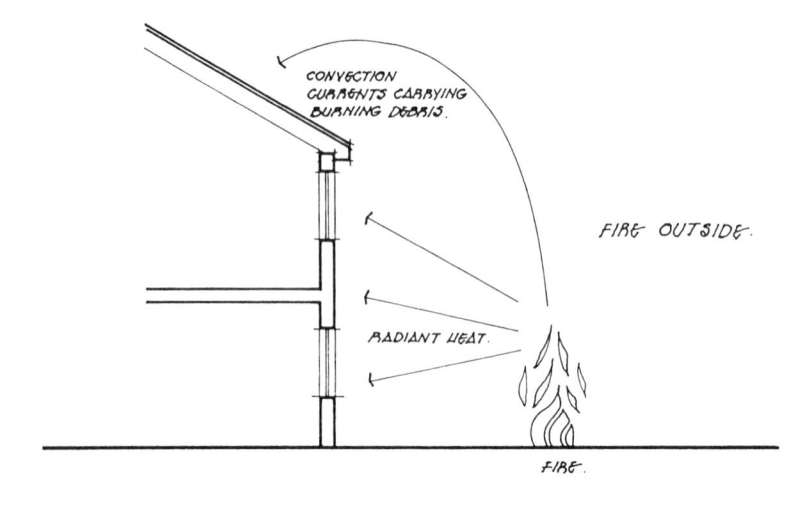

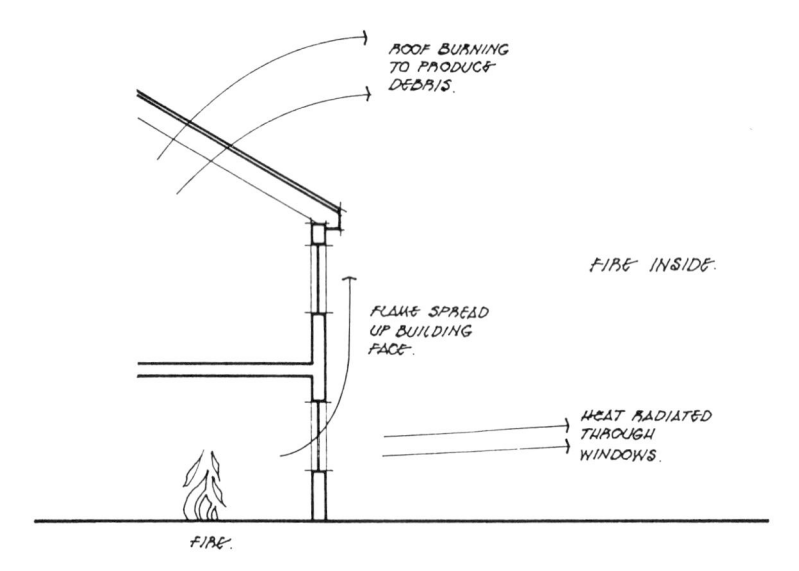

Figure 7.4 Envelope protection.

be considered, where this creates a risk for adjacent structures and where people are present. In such cases, the compartmentation standards go beyond the immediate needs of the occupants in the building.

If, however, the need is for property protection, the subdivision will be related to the level of acceptable risk and the zone boundaries have to be able to resist fully the complete burnout of the contents or until the fire is controlled. This concept accepts virtual complete loss in the fire zone but none in the adjoining areas. The basis of the subdivision may be the value of the contents, special risk areas or the natural division on the basis of activity carried out. The presence of additional fire safety measures, such as sprinkler installation, can justify adjustment in the size of the compartment or the fire resistance of the boundaries. The presence of a fire detection/alarm system can sometimes be credited with benefit, provided that it ensures a prompt attack on the fire.

The prevention of fire spread beyond the original occupancy or building into an adjacent property or a separately owned part of the building is a legal recognition for protection against the folly of another person and has been a component of fire regulations. Separating walls between properties adjoining each other, or the presence of compartment walls and floors within the same building as with a block of flats, are traceable to this philosophy.

If the adjacent property is physically separated, the concept still applies, although the solutions require slightly different treatment (Figure 7.4). For small separation (i.e. 1 m or less), the external walls are considered to act as compartment walls between adjoining buildings. For larger separations, the requirements are varied according to the level of the risk, taking account of the separation distance, size of openings, fire resistance of the wall and the presence of combustible cladding (Figure 7.5). The risk is due to radiation from the emitted flames from the building on fire, flaming on the façade which subjects the adjacent building to radiant heat and burning brands. The risk is calculable and methods have been developed to assess the hazard from different building configurations.

There are facilities in buildings which provide communication between different compartments, either horizontally or vertically. Service ducts, means for ventilation, stairways and lift shafts are examples of such communication routes. Protection is needed to ensure that a fire will not be communicated through such means, thus bypassing the compartment boundaries. Often it is also necessary to protect some of these communication facilities, such as escape stairways, from the effects of a fire at any level, so that they become unusable by the occupants or the fire-fighters. This is achieved by considering these as separate compartments and providing enclosure boundaries of fire-resisting construction. All openings in such enclosures and other compartment boundaries need to be fully protected; in practice, inadequate treatment of these is often responsible for the transfer of a fire from one compartment to another.

7.4 ESSENTIAL COMPARTMENTATION

On the basis of the previous analysis it is possible to visualize situations

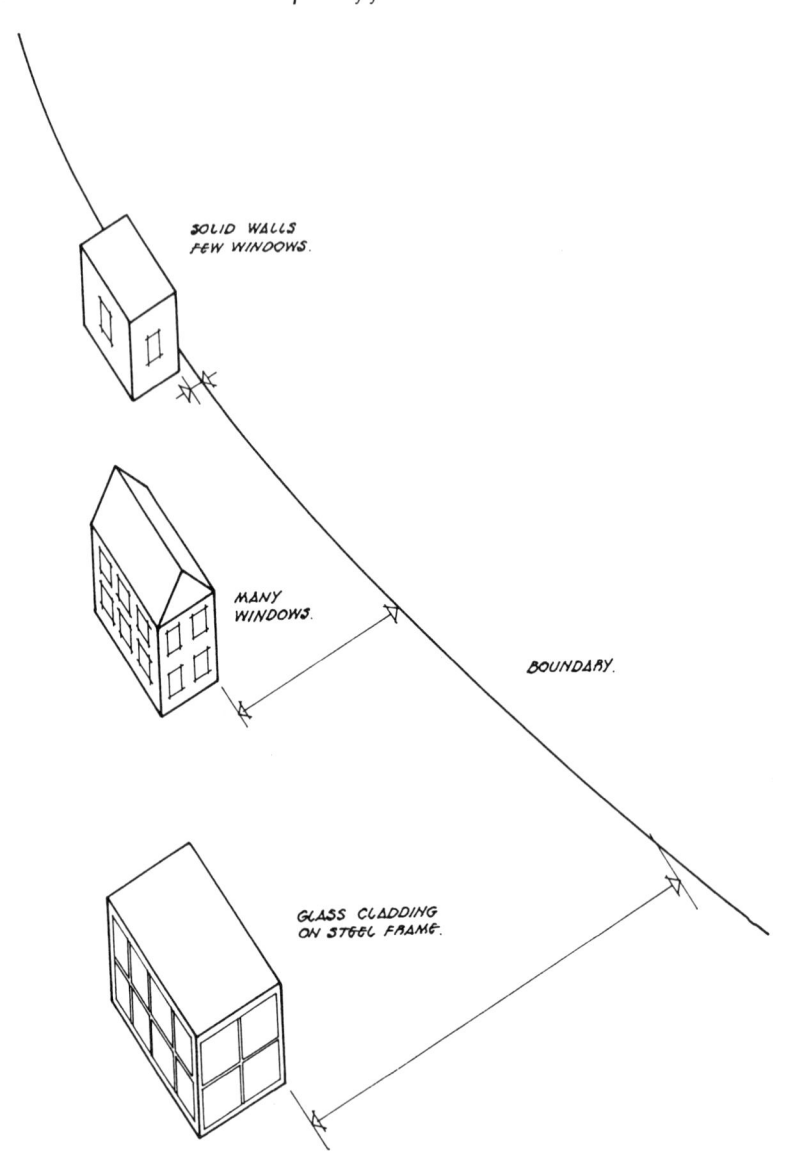

Figure 7.5 Distance to the boundary.

where compartmentation must be provided irrespective of the use of the building. These depend on the fire safety objective, as follows.

Life safety
- Duct and shafts linking compartments.
- Escape stairways, lift shafts, etc.

- Occupancies with special escape problems.
- Adjoining occupancies.
- Adjacent occupancies.
- Special risk areas.

Property protection
- Adjacent occupancies.
- Adjoining occupancies.
- Special risk areas.
- High loss value contents.

7.5 OPTIONAL COMPARTMENTATION

In addition to essential compartmentation, it may be considered desirable by the owner or the occupier of a building to provide additional compartmentation. This may be to achieve a higher standard of safety, to benefit from the reduction in fire resistance requirements for boundaries or to benefit from reductions in insurance premiums. Such additional measures are often not needed directly for life safety purposes, except where further subdivision facilitates the escape provisions. For property protection purposes, there can be distinct benefits in providing additional compartmentation as many insurance premiums are linked with an estimate of potential loss. There are no published rules in this connection and the terms have to be negotiated with the insurance companies. Under the regulations where concessions are made in fire resistance requirements for increased compartmentation, the benefits are in the constructional costs for the building.

7.6 FIRE RESISTANCE REQUIREMENTS

The compartment boundaries and other elements which are essential for the stability of these boundaries are required to possess a degree of fire resistance: this means that their design should be such that the fire of the severity expected in the compartment will not cause a rupture of the boundaries, nor be transferred to other parts, even if the boundaries remain stable. The determination of fire resistance is a well-established technology and has nearly a hundred years' history behind it and a vast store of knowledge on constructional specifications to achieve it.

The simplest solution for a designer is to look up the tables in the Building Regulations or the Approved Documents; these specify for each purpose group a range of fire resistance requirements in relation to the use and the physical characteristics of the building. Fire resistance requirements can vary from as little as 30 min to 120 min: the lowest requirement is for a single-storey building of low risk potential such as a house; the highest requirement is for buildings of large dimensions with heavy fire loads such as high-rise storage buildings. The height of a building is taken into consideration and demarcations are provided to distinguish among single-storey, low-rise, medium-rise and high-rise buildings. Traditionally high rise was considered to be the height beyond which external rescue was not possible – i.e. it was related to the rescue facilities. The 28 m (100 ft) ladder was a typical criterion, and although the height has now been increased to 30 m, still no consideration is given to tall monitors now used by the brigades. The intermediate divisions tend to be somewhat arbitrary; in the new Approved Document B in England and Wales [3] heights of 5 m and 20 m have been used as intermediate steps.

If the purpose of fire resistance requirements at one level is to ensure the safety of the occupants, this approach gives the impression that fires in high buildings are more severe than in low-rise ones and that fires in basements are more severe than above ground. However, neither of these statements is quite true; indeed the true reasoning is not known, except that instinctively authorities feel that larger buildings should have additional safety factors attached as the consequences of failure are more severe.

The expected fire severity for any compartment can be computed from a knowledge of the combustible contents (i.e the fuel) and its rate of burning. The burning rate is influenced by the nature of the fuel and the availability of air which supplies oxygen for combustion purposes. The thermal characteristics of the boundaries also have some influence as an insulating construction will retain more heat within the environment, thereby leading to high ambient temperatures. Simple formulae have been devised which allow such calculations to be made. If this is done, it will be found that most normal office buildings just about manage to qualify for fire severity which corresponds to a 60 min fire resistance. This means that if fires were allowed to burn freely, most small buildings may collapse but most other buildings though damaged will remain intact. In practice, of course the Fire Brigade will prevent free burning fires, and statistics show that most buildings are not subjected to full fire severity – the proportion may be as high as 90%.

If life safety were the primary concern, then the fire resistance requirements could be much lower than they are at present. Even for phased evacuation, the time for which a building needs to be protected can be less than 60 min. However, if it were considered necessary to protect the neighbouring properties from damage or to provide facilities for fire-fighting (as in the latest Approved Document), then an additional safety factor needs to be applied in those buildings where such considerations have to be applied.

A factor which has a significant influence on the likely fire severity is the existence of an automatic fire extinguishing system such as a sprinkler installation. The system is designed to operate while the fire is still small and either to extinguish it or keep it under control. This means that in buildings where properly designed and maintained sprinkler installations are provided the fire severity will be considerably smaller than expected from a consideration of the fire load and its rate of burning. The new Approved Document for England and Wales has given a general consideration to this provision for a range of non-residential buildings, with relaxation of fire resistance and compartment size requirements. The positive role of sprinkler systems has been recognized to such an extent that high-rise non-residential buildings (>30 m height) require a sprinkler system as an obligatory requirement.

7.7 FIRE RESISTANCE PROVISION

The provision of fire resistance requires the designer to establish the necessary level and then to select or design the construction capable of providing the required fire resistance. It needs to be borne in mind that although fire resistance is expressed in time, this time is not the same as the time for which a fire may be expected to last. The fire resistance time specified for buildings is an experimental time based on an estimate of the fire load and its burning characteristics, adjusted if necessary for other factors. Similarly, the fire resistance of a construction is determined by subjecting it to a laboratory test under specified exposure conditions. The actual fire duration and the heat severity are different from fire to fire, even in similar occupancies. The confusion arises because time is used as a common parameter in both cases. Confusion might be avoided if the expression 'fire resistance' was qualified in some way; for example, the following expressions may be used:

Design fire resistance – the period considered necessary for the protection of a building from considerations of the fire load, occupancy factors and safety needs.
Experimental fire resistance – the time for which an element is able to satisfy the relevant criteria of the standard test in order to satisfy the design fire resistance needs.
Calculated fire resistance – using a calculation procedure to establish the time for which an element is considered capable of satisfying the fire resistance criteria in lieu of subjecting it to a test.

The Building Regulations specify the first type of fire resistance, the laboratories establish the second type and the design engineers compute the third. Many publications provide guidance on the experimental and calculated fire resistance; the manufacturers of certain materials and systems publish data on their products; and the material or design codes include tables and procedures for the calculated fire resistance. A fire safety engineer can be expected to have knowledge of all three approaches.

7.8 AVAILABLE TECHNOLOGIES

At present no quantifiable techniques are available to establish the necessary size of a compartment on the basis of estimating the risk, except for the space separation between adjacent buildings. The allowable compartment sizes or the needs for essential compartmentation are specified in Building Regulations or the Approved Documents. The system has recently been reviewed with the publication of the new version of the Approved Document B on fire safety for England and Wales. This document includes new ideas and some of the important ones in connection with fire containment are:

1. Removal of linkage between fire resistance and compartment sizes.
2. Some rationalization of compartment sizes, their expression in terms of floor area for all buildings, except for storage occupancies where limiting volumes are specified.
3. Exemption of certain buildings from compartment size limitation – e.g. single storey buildings, multi-storey residential buildings, except those of an institutional type; in residential building natural compartmentation occurs at the boundaries of different occupancies.
4. Increase of compartment sizes, by 100%, if a sprinkler system is provided.
5. Obligation to design all floors as compartment floors in high-rise buildings and certain other buildings posing escape problems.

For different buildings occupancies these new concepts may be summarized as:

- Residential, institutional and hotels, etc.: all floors are compartment floors, and there are requirements, or concessions, for a sprinkler system.
- Office, shop, assembly, industrial and storage buildings; limited in height to 30 m without sprinklers; with sprinklers, the maximum fire resistance requirement is 120 min.
- Compartment size limitation: none for offices; shops and assembly buildings, a limit of 2000 m^2 floor area without sprinklers and 4000 m^2 with sprinklers. Industrial buildings have a size limitation of 7000 m^2 up to 20 m height and 2000 m^2 above that without sprinklers; these sizes can be doubled with sprinklers. Storage buildings have a size limitation of 2000 m^3 up to 20 m height and 4000 m^3 above that without sprinklers; these sizes can be doubled with sprinklers.

7.9 FUTURE NEEDS

There is a pressing need for a detailed study on the logic and rationale of compartment size limitations, the development of a quantifiable basis for determining the size and relating the compartmentation requirements with fire safety objectives. Such a study is essential if fire safety engineering techniques are to be applied to compartmentation requirements, and if these are to be made a part of the fire safety technology. There is no doubt that compartmentation is an essential part of a fire safety system but it needs to be based on quantifiable relationships and the current historical systems need to be subjected to an analytical scrutiny.

REFERENCES

1. HMSO (1946) *Fire Grading of Buildings: Part I, General Principles and Structural Precautions*, Post-War Building Studies No. 28, HMSO, London.
2. Malhotra, H. L. (1988) *Fire Safety in Buildings*, Building Research Establishment, Borehamwood, December.
3. HMSO (1992) *Approved Document B: Fire Safety*, The Building Regulations, HMSO, London.

8
Smoke control in shopping malls and atria

Howard Morgan

8.1 INTRODUCTION

Fire protection measures have historically been used in large buildings to minimize the consequences of serious fires, both in terms of property loss and of life loss. In this chapter emphasis will be placed on life safety, but all the principles discussed can also be applied to property protection.

In the UK (this chapter will concentrate on practices in the UK) protection has been largely based on the provision of fire-resisting construction to divide large buildings into manageable volumes, called fire compartments, to reduce the space affected by the fire, coupled with means of escape rules to ensure safe evacuation of people.

Means of escape standards typically specify maximum travel distance to an exit and the minimum widths of the exits. They also specify that routes (e.g. stairways) should be provided, protected from the rest of building by fire-resisting construction, for evacuees to leave the building to a place of safety without being threatened by a fire in any other fire compartment which they need to pass on the way out.

The UK Building Regulations have always contained a provision for innovation and for exceptional designs, via the Relaxation Procedures. These procedures can be summarized as allowing the designer to depart from the set rules, provided that he can prove to the relevant regulatory authority that he can achieve at least the same level of safety by other means. Every building must be considered as a whole by assessing the total package of safety measures.

The foregoing is particularly important because it is usually impossible to build a shopping complex or atrium building while keeping all fire compartments within the prescribed volume limits. In the UK a relaxation of the Building Regulations is usually required for all such structures.

Covered shopping complexes or malls present an additional problem,

A brief history of smoke ventilation

in that the covered malls take the place of open-sky streets. A covered mall cannot be a place of safety in the same way that an open-sky street is. The problem is resolved in the UK by:

1. specifying the means of escape within each shop unit as though the mall was indeed a place of safety – see e.g. [1];
2. recognizing that the mall is an additional stage on the escape route to an actual place of safety outside the building;
3. requiring special measures to protect escapees using the malls – see e.g. [2].

In the UK, smoke ventilation is a crucial part of the package of measures used to protect the mall. It is also commonly used as part of the package of measures adopted for atrium buildings to compensate for the 'excess volume' of the undivided fire compartment created by the presence of the atrium.

8.2 A BRIEF HISTORY OF SMOKE VENTILATION

Smoke ventilation is nothing new: our distant ancestors knew that they needed a hole in the roof of a hut if they wanted to light a fire inside, otherwise the occupants would be choked by smoke. Modern smoke ventilation merely applies the same principle to large fires in modern buildings.

Smoke ventilation as a dedicated fire precaution became popular for industrial buildings following some large fires – e.g. General Motors plant in Michigan, USA in 1953; the Jaguar car plant in Coventry in 1957; and the Vauxhall Motors plant at Luton in 1963. (Only the last of these three had automatic ventilators [3].) During the 1960s, the Fire Research Station in the UK developed design algorithms suitable for circumstances where the fire would be directly below the thermally buoyant smoke layer formed beneath the ceiling [4,5]; the technique was mostly used as a way of reducing property damage by allowing fire-fighting to become much more effective.

A fire in the linked Wulfrun and Mander Shopping Centres [6] alerted people to the very considerable potential for spread of smoky gases in covered malls. It was realized that such a fire could cause large losses of life if it occurred when the mall was being used by the public.

It was realized that the smoke ventilation approach already developed for large spaces could be adapted to keep smoke entering a mall safely above people's heads, thus protecting the means of escape in the mall. Research on the way in which smoke moves within malls continued through the 1970s, leading to the development of design formulae for calculating the movement of the smoke, and its mixing with air, and hence the sizes of the vents or fans needed to exhaust the smoky gases in order to maintain the smoke layer in the malls at a safe height. A summary of the design advice available from the Fire Research Station was published in 1979 [7]. This advice has been expanded and updated in the light of further research and experience. [8]

In the late 1970s research began on the related problems of atrium buildings, where the main feature is that of a central void rising through two or more storeys, allowing any smoke entering the void to affect more storeys than the original fire storey, unless of course these floors are separated from the atrium by fire-resisting construction – in which case the atrium is merely a room with an unusually high ceiling! This latter case is trivial and need not concern us further. The remainder of this chapter will discuss only atria whose adjacent storeys are not so separated from the void.

It should be readily obvious that a shopping mall of two or three storeys represents a special case of atrium: it is an atrium with a single class of occupancy. The smoke movement will be similar, the smoke hazards will be similar and the smoke control solutions can be expected to be similar.

The pivotal problem for both malls and atria is that smoke entering the void must not be allowed to endanger safe escape for people in the mall or the atrium itself, or for people in any adjacent space open to the mall or atrium on any storey.

In atria or multi-storey malls every storey open to the void is potentially rapidly affected by smoke from a fire on any other storey. Two fires that illustrate this were the fire in the Regency Hyatt Hotel at O'Hare in Chicago [9], and the fire in the St John's Centre in Liverpool [10]. It follows that to protect the safe escape of the building's occupants special measures are needed for atria as for malls. Any measures to protect egress will also assist easier entry for fire-fighters. Hence the same measures will serve to improve the property protection aspects of the fire protection package.

8.3 BASIC PRINCIPLES OF SMOKE VENTILATION

Smoke ventilation is used when the fire is in the same space as the people, contents or escape routes being protected. Air mixes into fire/smoke plumes as they rise, increasing the total volume of smoky gases (Figure 8.1). These gases flow outward below the ceiling until they reach a barrier (e.g. the walls or a downstand).

The gases then form a deepening layer, whose buoyancy can drive smoky gases through natural vents (Figure 8.2). Alternatively, smoky gases can be removed from the layer by using fans. For any given size of fire, an equilibrium can be reached where the quantity of gases being removed equals the quantity entering the layer in the plume – no significant amount of air mixes upwards into the base of the buoyant

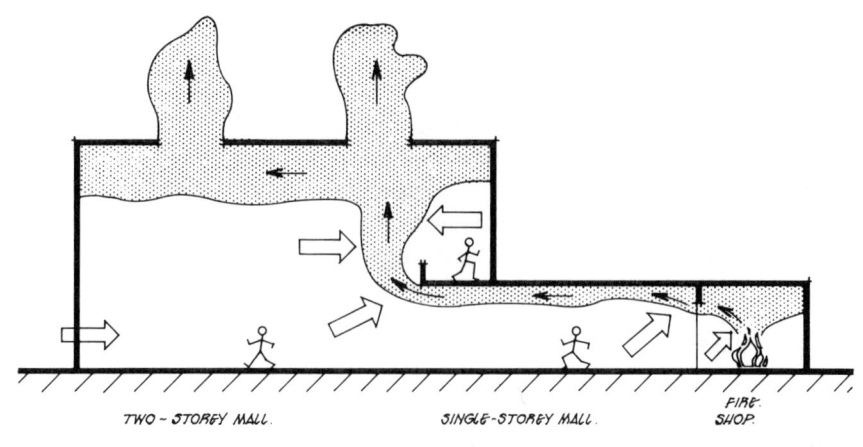

Figure 8.1 Smoke spread and main entrainment sites in single- and two-storey malls.

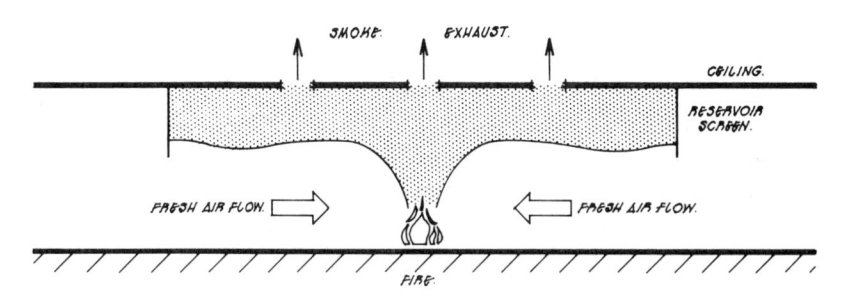

Figure 8.2 Principles of system needed to contain smoke in a well-defined layer (section along mall).

smoke layer. Sufficient clear air must enter the space below the layer to replace the gases being removed from the layer, otherwise the smoke ventilation system, will not work.

The capacity of the smoke exhaust system depends primarily on the size of the fire assumed for design and on the height through which the hot gases have to rise before entering the smoke layer. This height is usually chosen to achieve safety – the smoke must be above people's heads on the highest exposed escape route.

The fire size is more difficult to assess. Using an 'average' fire implies that one is designing into the system a 50% failure rate. Yet there is no limit to how large fires can grow. Since larger fires occur less often than smaller fires, the best approach is a statistical one, whereby one tries to find an acceptable 'plausible but pessimistic' size. Other fire precautions (especially the use of sprinklers) play a major role in reducing the acceptable size of fire for design. It has to be admitted, however, that this crucial step of choosing a design fire can often only be done by experienced assessment wherever a lack of data prevents more rigorous methods being used. Note that a design fire of 3 m × 3 m, having a convective heat flow of 5 MW, has been adopted since 1972 in the UK for retail premises having sprinklers and opening onto covered malls.

In a fire most of the important factors are time dependent. The time for the fire to grow from ignition to the design fire size, for the mall to fill with smoke and for people to evacuate the malls are all-important and can, in principle, be used in designing smoke control systems. Currently there are no reliable data on fire growth rates in retail premises. The time needed for safe evacuation is usually unquantifiable, but can be significantly longer than the 2.5 min often used in the UK to calculate the minimum widths needed for exits and stairs, etc. Consequently, it is still unusual to base a design on a growing fire. It remains more common to use 'plausibly pessimistic' steady fires for design in view of the greater ease with which they can be assessed.

8.4 DESIGN PARAMETERS FOR ATRIA AND MALLS

In this section we will discuss principles or applications which have particular relevance to atria and malls, with reference to other documents which can be used for more detailed design. Most of these describe procedures developed by the Fire Research Station in the UK, but designers should be aware of design procedures developed in the USA for NFPA [11]. It is also possible to use computational fluid dynamics methods when designing for specific buildings, although this lies outside the scope of the present chapter.

Fires on the atrium floor

A fire on the atrium floor is no different in principle to the simple geometry case of a large room as far as mixing into the plume is concerned. Calculation is straightforward [4,5,12], although tall narrow atria can become smokelogged when the rising plume expands to fill the atrium cross-section, even before reaching the design layer base [11].

Fires in rooms adjacent to the void

A fire in an adjacent space is more complicated to calculate than one on the atrium floor. Fully involved fires which fill the adjacent space are particularly difficult, and should not be allowed in the design unless the rooms are small and compartmented from each other by fire-resisting construction. Sprinklers are useful in limiting design fire sizes in larger rooms, and can usually be specified wherever the design fire would otherwise be impracticably large. They can often therefore form an essential part of the smoke control system, and the design of the sprinkler system should be integrated with the design of the smoke ventilation. All too often the two systems are in fact designed independently, sometimes to the disadvantage of both. Where the fire is smaller than the room's floor area, the fire is usually fuel-bed controlled. The quantity of smoky gases entering the atrium or mall can be calculated as the 'balance-point' between the mixing of air into the plume inside the room and the outflow of gases through the openings to the atrium or mall. Plume entrainment in the room depends on the size of fire and on whether or not air can approach the plume freely from all directions [13]. A symmetrical approach flow is more likely in a large room with wide openings, and an asymmetrical approach flow is more likely in small rooms with limited width openings. Entrainment can be almost twice as great in the latter case as in the former. Hansell has developed design procedures incorporating these phenomena [14]. There is also a considerable complexity of theoretical interpretation concerning the calculation of flows out of small openings when the fire is close to flashover. Simpler design formulae have been developed for fires in large compartments having wide openings [15,16], fortunately a fairly common design scenario. In the UK all this complexity has long been bypassed for shopping malls by specifying a 5 MW, 3 m × 3 m fire for shops having sprinklers [8].

Canopies or balconies outside the room opening

The next stage in calculating mixing of air into the smoke occurs where the gases rise from the opening to a canopy or balcony projecting beyond the opening, or into the layer formed beneath the ceiling of a single-storey mall. The former case can be approximated by a simple doubling of the mass flow rate leaving the room opening [15], or it can be calculated more precisely using empirical correlations [14]. The latter case has been found empirically to be calculable by a simple approximation for single-storey malls. This can be done by pretending that the fire in the shop is actually in the mall and then doubling the entrainment predicted by the usual large-fire plume equation. This is the approach usually followed in the UK [7,8]. Note, however, that other empirical correlations are possible for the same scenario [17]. None of these correlations can be expected to apply if the height of rise of the plume outside the opening is large: a 'spill plume' calculation is needed instead (see below).

Spill plumes in the atrium or mall

When the smoky gases enter the atrium or mall void, they rise as a plume into the buoyant layer formed beneath the atrium (or mall) ceiling

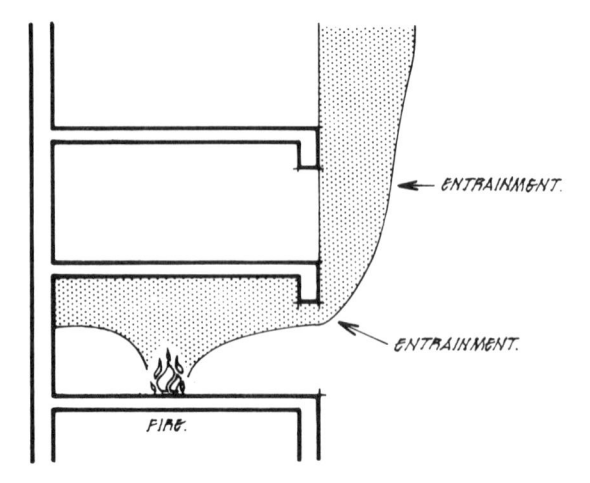

Figure 8.3 Adhered plume.

(Figures 8.3 and 8.4). This spill plume is usually longer than it is broad, depending on the length of the edge past which it spilled. For this reason, such plumes are often called line or two-dimensional plumes. These plumes entrain large amounts of air, which must be calculated in order to arrive at the total quantity of gases entering the smoke layer (and hence the smoke exhaust capacity needed). Plumes rising up a vertical surface (Figure 8.3) entrain much less air than a free plume rising well away from the walls (Figure 8.4). Approximately only a quarter as much air mixes

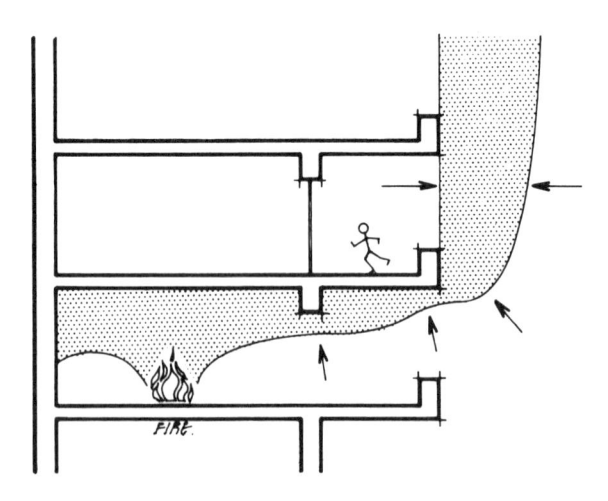

Figure 8.4 Free plume (arrows indicate main points of entrainment).

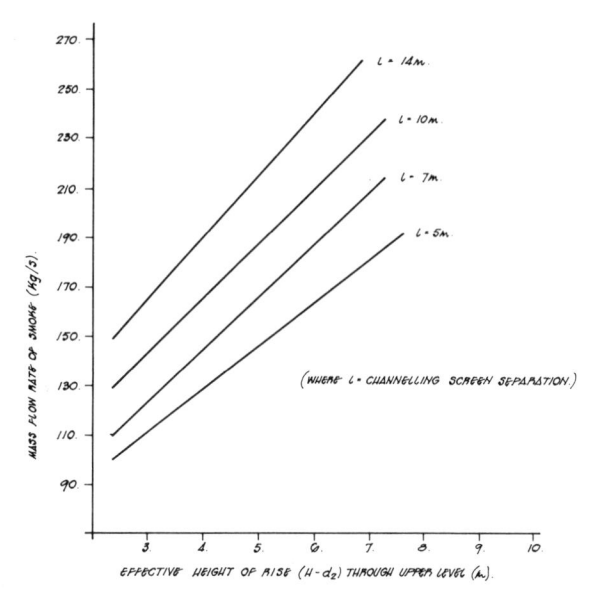

Figure 8.5 Mass of smoke entering ceiling reservoir per second from a 5 MW fire – large voids.

into the plume per unit height. This can be expressed as a halving of the entrainment constant used in the adhered (or single-sided) plume calculations set out by Morgan and Hansell [15]. More recent work [18] has confirmed Grella and Faeth's earlier work on entrainment into adhered plumes [19], and has confirmed that the Morgan and Hansell method for adhered plumes works with the smaller value of the entrainment constant. It is hoped that a document will be published in the near future summarizing the design approach being developed by Hansell and Morgan. Entrainment into free plumes can be calculated also [8,15]. Morgan and Gardner [8] include pre-calculated graphs for those users who prefer not to solve the equations for themselves. Figures 8.5 and 8.6 show examples of such results for a mall and an 'office atrium' respectively. Other approaches using simpler correlations are possible – see, e.g., Thomas [20]. Most such correlations incorporate all the fire-room and under-balcony entrainment within the spill plume correlation. Hence they are more specific to the original experimental geometry than the more complicated procedure outlined above.

Limits to smoke ventilation of atria and malls

It can be seen from both Figures 8.5 and 8.6 that the mass flow rate of smoky gases increases rapidly with height. This can, to some extent, be

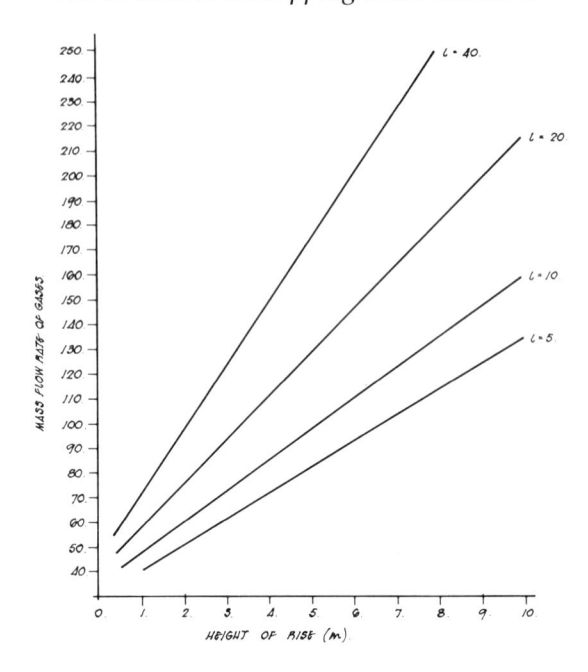

Figure 8.6 Free plume from open-plan sprinklered office (heat output: 1 MW; downstand depths at opening: 1.0 m.

offset by channelling the gases approaching the void edge before they spill into the void, in order to make the spill plume shorter and more compact. Unfortunately, even for compact plumes, there rapidly comes a limit where the quantity of smoke entering the layer becomes impractical to exhaust. Another limit can occur with smaller fires, where the temperature in the buoyant layer is too low for smoke to remain stably stratified in the presence of draughts in the atrium or mall. The critical temperature is not known, but a lower limit of 20°C is often adopted in the UK. It follows that there is an upper limit to the number of storeys that can reasonably be protected by smoke ventilation if they are open to the atrium or mall. For atria, that limit seems to be where the smoke base is no higher than between 8 m and 18 m above the fire floor (depending on the building geometry). Malls in the UK are not allowed to have more than two storeys open to the common void through which smoke can rise.

Alternatives to atrium/mall smoke ventilation

There is, in principle, no limit to the number of storeys which can be open to the atrium or mall, provided that smoke from all except the topmost storeys (which are subject to the above limits) is prevented from spilling into the atrium or mall; there is considerable scope for creative fire engineering here. Storeys which are only linked to the atrium by relatively small 'leakage' openings can be protected by a different form of smoke control. Pressure differences can be created across those openings, either by raising the pressure in the adjacent rooms or by reducing the pressure in the atrium itself. The buoyancy of the fire gases themselves can be taken advantage of in achieving these pressure differences.

Smoke exhaust

Exhaust of smoke from the atrium or mall smoke reservoir depends on the quantity of smoky gas, its temperature and the layer depth. Design procedures are effectively the same for malls and atria as for industrial premises. Details can be found elsewhere [4,5,7,8,12,15]. Either powered or natural ventilators can be used – although never together in the same smoke reservoir. Both types of ventilator have their advantages and disadvantages, and the choice should be based on the particular circumstances in which they will operate. For example, natural ventilators should never be used where they could be subject to wind-induced overpressures; fans require protected back-up power supplies which may be difficult to provide.

8.5 ACTIVATION OF THE SYSTEM

Where smoke control systems are provided for the primary purpose of protecting life safety, all parts of the system must come into operation since having a human being in the 'decision process' introduces too good an opportunity for confusion and delay. Automatic smoke detection causing an automatic initiation of smoke exhaust is a particularly effective combination in most circumstances. It is also desirable to include in the design some provision for override facilities which can be operated by the fire services to assist them during fire-fighting operations.

ACKNOWLEDGEMENT

This chapter is part of the work of the Fire Research Station, Building Research Establishment Agency, Department of the Environment. It is reproduced by permission of the Controller of the HMSO: Crown Copyright 1991.

REFERENCES

1. British Standards Institution (1985) *Fire Precautions in the Design and Construction of Buildings, BS 5588, Part 2: Code of Practice for Shops*, BSI, London.

2. British Standards Institution (1991) *Fire Precautions in the Design and Construction of Buildings, BS 5588, Part 10: Code of Practice for Shopping Malls*, BSI, London.
3. Thomas, P. and Hinkley, P. L. (1964) Roof venting theory and the Vauxhall fire. *Fire Protection Review*, 27, 282, 208–9.
4. Thomas, P., Hinkley, P. L., Theobald, C. R. and Simms, D. L. (1963) *Investigations into the Flow of Hot Gases in Roof Venting*. Fire Research Technical Paper No. 7, HMSO, London.
5. Thomas, P. and Hinkley, P. L. (1964) *Design of Roof-venting Systems for Single-storey Buildings*, Fire Research Technical Paper No. 10, HMSO, London.
6. Silcock, A. and Hinkley, P. L. (1971) *Fire at Wulfrun Shopping Centre, Wolverhampton*, Fire Research Station Fire Research Note 878, BRE, Borehamwood.
7. Morgan, H. (1979) *Smoke Control Methods in Closed Shopping Centres of One or More Storeys: A Design Summary*, Building Research Establishment Report, HMSO, London.
8. Morgan, H. and Gardner, J. P. (1990) *Design Principles for Smoke Ventilation in Enclosed Shopping Centres*, Building Research Establishment Report, BR 186, BRE, Garston.
9. Sherry, J. A. (1973) An atrium fire. *Fire Journal*, **67**(6), 39–41.
10. Morgan, H. and Savage, N. P. (1980) *A Study of a Large Fire in a Covered Shopping Complex: St John's Centre 1977*, Building Research Establishment Current Paper CP 10/80, BRE, Borehamwood.
11. Milks, J. (1990) Smoke management for covered malls and atria. *Fire Technology*, **26**(3), 223–43.
12. Morgan, H. (1985) *A Simplified Approach to Smoke-Ventilation Calculations*, Building Research Establishment Information Paper, IP 19/85, BRE, Borehamwood.
13. Zukoski, E. E., Kubota, T., and Cetegan, B. (1981) Entrainment in fire plumes. *Fire Safety Journal*, 3, 107ff.
14. Hansell, G. O. (forthcoming) Heat and mass transfer processes affecting smoke control in atrium buildings. PhD thesis.
15. Morgan, H. P. and Hansell, G. O. (1987) Atrium buildings: calculating smoke flows in atria for smoke control design. *Fire Safety Journal*, 12, 9–35.
16. Morgan, H. P. and Hansell, G. O., Atrium buildings smoke flows. *Fire Safety Journal*, **13**(2–3), 221–4.
17. Law, M. (1989) Design Formulae for Hot Gas Flow from Narrow Openings – Points for Consideration. Society of Fire Safety Engineers 'Flow through Openings' Meeting, Borehamwood, 13 June.
18. Hansell, G. O., Morgan, H. P. and Marshall, N. R. (forthcoming) *Smoke Flow Experiments in a Model Atrium*.
19. Grella, J. J. and Faeth, G. M. (1975) Measurements in a two-dimensional thermal plume along a vertical adiabatic wall. *Journal of Fluid Mechanics*, **71**(4), 701–10.
20. Thomas, P. (1987) On the upward movement of smoke and related shopping mall problems. *Fire Safety Journal*, 12, 191–203.

9
Fire information

Paul Stollard

9.1 INTRODUCTION

This chapter is intended to guide both architects and the staff of statutory authorities to all the information they may need in working from first principles. There is so much information available that there is a real danger that important issues can become lost in a plethora of insignificant and confusing detail. This chapter does not attempt to summarize all this information; instead it should serve as a guide to that which is most important for those working within the UK.

The next three sections briefly outline the legislation relevant to the different parts of the United Kingdom. The relationship between Acts and regulations and codes is examined and the application and authority of the various documents is outlined. The precise details of the legislation are not reproduced as designers must have their own copies of the Approved Documents or Building Regulations, but an attempt is made to explain the relevance and application of the different parts of the legislation.

Much of the legislation refers directly to British Standards, and these are introduced in section 9.4. These documents occupy a curious position, almost akin to legislation in their authority, and it is essential for the designer to be aware of the most important standards and of their numbers and codes.

The next section summarizes all the other guidance available from research, governmental and trade bodies. This includes a review of other textbooks which are available. A list of advice sources also available is included in the final section.

References will not be listed at the end of this chapter, but given in full under the relevant section, so that this chapter can serve as a simple index to designers and the statutory authorities seeking specific guidance.

9.2 LEGISLATION IN GREAT BRITAIN

Most of the British legislation relating to fire safety has been enacted as a result of tragedies or particular fires. It has therefore developed in a piecemeal fashion, with different sets of legislation referring to different building types, and distinguishing between new buildings and existing buildings. This incremental process has left some gaps and created areas of overlap; in addition, it can be very confusing and is not always consistent or logical. This is another reason why the designer must consider fire safety from principles and not from legislative compliance. Although the development of the legislation may be of only passing interest to the designer, various books give a good historical perspective of fire safety law, notably R. E. H. Read and W. A. Morris, *Aspects of Fire Precautions in Buildings* (BRE, 1988).

It is unlikely that any major overhaul of the system will be undertaken and the process of refinement and replacement will continue, giving the law an almost 'geological' quality, with Act and regulation superseding previous Act and regulation – sometimes completely as in a consolidating Act and sometimes only partially. The most recent review of fire and building regulations undertaken for the Enterprise Deregulation Unit of the Department of Trade and Industry in conjunction with the Home Office and the Department of the Environment by Bickerdike Allen and Partners (HMSO, 1990) advocated only minor changes and improvements.

Not all documents have the same status and it is important for the architect to be aware that many of those, to which statutory authorities may demand slavish obedience, are not in fact binding, but rather only advisory. Legal requirements can be categorized in order of their significance, as follows:

1. Acts of Parliament (or Enactments) – these are the laws passed by Parliament which have to be obeyed; however, these very rarely contain technical details and normally only set up procedures or empower government departments to make regulations.
2. Regulations (or Statutory Instruments) – these are made by government departments under enabling Acts and have the force of law; they used to contain technical information, but in the most recent Building Regulations for England and Wales (1992) the trend has been to move towards functional requirements outlining requirements in general terms.
3. Approved Documents and guides – these do not constitute the law, rather they are normally advisory documents which suggest one method of fulfilling the requirements of the regulations.

British legislation can be divided into two broad categories: that dealing with new buildings, and that dealing with existing buildings. New

Legislation in Great Britain

buildings are primarily the responsibility of the building control departments of local authorities, while existing buildings are primarily the responsibility of the fire prevention departments of fire authorities. The first two areas to be considered are new buildings in England and Wales, and in Scotland – separate systems having been retained in different countries. Then existing buildings which are covered by the Fire Precautions Act 1971 (and which covers the whole of the mainland) is considered. Following this, there is coverage of the Health and Safety at Work Act 1974 which makes particular regulations about places of work, both new and existing. Next the Licensing Acts, which again cover new and existing premises, are outlined; these nearly always involve the fire authorities in advising the licensing justices. Finally in this section we outline other legislation which applies to specific local area or specific building types.

One of the problems for the architect is the sheer mass and complexity of the legislation which could apply; this section does not attempt to summarize the legislation, but rather describes the patchwork quilt of overlapping documentation, so that the design team can see how the different pieces interrelate.

The Building Regulations (England and Wales)

The Building Act 1984 consolidated the law relating to building control for England and Wales, but it contains no design information as such. The Act empowers the Secretary of State to make Building Regulations and under it the successive regulations have been issued. The Act also gives local authorities some detailed powers to deal with such matters as adequate drainage, building on filled ground and demolition. In addition, the Act gives local authorities the power (sections 24, 71 and 72) to demand the provision of means of escape in cases of certain buildings after consultation with the fire authority. This provision is designed to cover non-domestic buildings where the Building Regulations might not apply.

The current Building Regulations for England and Wales were made under the Building Act 1984, and Part B covering Fire Safety came into force in its present form on 1 June 1992; there are five functional requirements which apply to all new buildings (except prisons).

B1: Means of Escape
The building shall be designed and constructed so that there are means of escape in the case of fire from the building to a place of safety outside the building capable of being safely and effectively used at all material times.

B2: Internal Fire Spread (Linings)
(1) In order to inhibit the spread of fire within the building, the internal linings shall –

(a) resist the spread of flame over their surfaces; and

(b) have, if ignited, a rate of heat release which is reasonable in the circumstances.

(2) In this paragraph 'internal linings' means the materials lining any partition, wall, ceiling or other internal structure.

B3: *Internal Fire Spread (Structure)*

(1) The building shall be so constructed that, in the event of fire, its stability will be maintained for a reasonable period.

(2) A wall common to two or more buildings shall be designed and constructed so that it resists the spread of fire between those buildings. For the purposes of this sub-paragraph a house in a terrace and a semi-detached house are each to be treated as a separate building.

(3) To inhibit the spread of fire within the building, it shall be sub-divided with fire resisting construction to an extent appropriate to the size and intended use of the building.

(4) The building shall be designed and constructed so that the unseen spread of fire and smoke within concealed spaces in its structure and fabric is inhibited.

B4: *External Fire Spread*

(1) The external walls of the buildings shall resist the spread of fire over the walls and from one building to another, having regard to the height, use and position of the building.

(2) The roof of the building shall resist the spread of fire over the roof and from one building to another, having regard to the use and position of the building.

B5: *Access and Facilities for the Fire Service*

(1) The building shall be designed and constructed so as to provide facilities to assist fire fighters in the protection of life.

(2) Provision shall be made within the site of the building to enable fire appliances to gain access to the building.

Each of the five functional requirements states that the building should have a level of safety, but this is either undefined or described only as being 'reasonable'. The Building Regulations are supported by an Approved Document which defines what is the acceptable, or 'reasonable', level of safety through precise specification; this is: Department of the Environment and Welsh Office, *Approved Document B, Fire Safety, the Building Regulations 1991* (HMSO, 1991).

There is no obligation on the designer to follow the solutions outlined in the Approved Document and they are entitled to meet the relevant requirement in any other way which they can demonstrate offers an equivalent level of safety; however, virtually all architects and statutory authorities tend to treat the Approved Document as if it were gospel, and to treat alternative fire safety solutions with great scepticism. The position of Approved Documents can be compared with that of the *Highway Code*:

it is not an offence not to follow it, but should anything go wrong, this will be the standard which the courts would look to as representing reasonable practice. In reality, few architects have been brave enough to venture out on their own and the Approved Document has become almost a prescriptive piece of legislation which is obediently followed in virtually all designs.

The Building Regulations are administered by the local district councils. The architect can either go to them for approval or, in theory, to an Approved Inspector employed by the architect. In practice, there are no Approved Inspectors except for the National House Building Council, which acts only for its own members.

As the Building Regulations are now in the form of functional requirements, it is impossible to apply for relaxations.

The Approved Document is structured under five functional requirements with sections covering the major areas of importance.

B1: Means of Escape
Section 1: Dwelling house
Section 2: Flats and maisonettes
Section 3: Design for horizontal escape – buildings other than dwellings
Section 4: Design for vertical escape – buildings other than dwellings
Section 5: General provisions common to buildings other than dwellings.

B2: Internal Fire Spread (Linings)
Section 6: Wall and ceiling linings.

B3: Internal Fire Spread (Structure)
Section 7: Loadbearing elements of structure
Section 8: Compartmentation
Section 9: Concealed spaces (cavities)
Section 10: Protection of openings and fire stopping
Section 11: Special provisions for car parks and shopping complexes.

B4: External Fire Spread
Section 12: Construction of external walls
Section 13: Space separation
Section 14: Roof coverings.

B5: Access and Facilities for the Fire Service
Section 15: Fire mains
Section 16: Vehicle access
Section 17: Personnel access
Section 18: Venting of heat and smoke from basements.

Until the advent of the 1985 Building Regulations the 11 inner London boroughs and the City of London retained their own legislative requirements which traced their descent from laws first enacted in the twelfth century. The new Building Regulations were applied to inner London as from the 6 January 1986 and are now administered by the individual London boroughs. A number of specific regulations were, however, retained and still apply only to inner London, probably the most significant of these being the old sections 20 and 21 which set additional fire safety standards for tall buildings.

Building Standards (Scotland) Regulations

Scotland has always maintained its own set of Building Regulations, and at times they have been in advance of those of England and Wales. The set currently in force were produced in 1990 and, sadly, the opportunity was not taken to harmonize them completely with the Approved Document B which was then being prepared.

Like the Building Regulations for England and Wales, the actual regulations for Scotland consist of a simple series of functional requirements. The two relating to fire safety are as follows.

12. Structural Fire Precautions
Every building shall be so constructed that for a reasonable period, in the event of a fire –
  (a) its stability is maintained
  (b) the spread of fire and smoke within the building is inhibited
  (c) the spread of fire to and from other buildings is inhibited.

13. Means of Escape from Fire and Facilities for Fire Fighting
Every building shall be provided with –
  (a) adequate means of escape in the event of fire; and
  (b) adequate fire fighting facilities.

These two regulations can be complied with by following the Technical Standards which accompany the Building Standards Regulations, or by conforming with the demand to satisfy provisions or by any other measures which the designers can show satisfy the standards. The relevant Technical Standards are Part D or structural fire precautions and part E for means of escape and facilities for fire-fighting. Although there is in theory the option for the designer to offer an alternative fire engineering solution, the emphasis on the Technical Standards deters many. The presentation of the Technical Standards, together with the Regulations, in a single binder, and deemed to satisfy provisions only encourages the statutory authorities to treat the Technical Standards as if they were in fact the Regulations.

Home Office guides, Fire Precautions Act 1971

Unlike the various sets of Building Regulations, the Fire Precautions Act 1971 applies to both new and existing buildings. The Act permits the Home Secretary to designate building types where Fire Certificates must be obtained from the local fire authority, and since the introduction of the Act three designation orders have been made. The first, in 1972, covered hotels and boarding houses; the other two were in 1977 and covered, first, factories, and secondly, offices, shops and railway premises.

For the designated building types, the Home Office and the Scottish Home and Health Department have published guidance on the standards that are required for certification. The guidance for hotels and boarding-houses is: *Guide to the Fire Precautions Act 1971: 1, Hotels and Boarding Houses* (HMSO, 1972). This guide covers means of escape, walls and ceiling finishes on escape routes, fittings and furniture on escape routes, fire-warning equipment and fire-fighting equipment.

The guide for factories, offices, shops and railway premises is: *Guide to Fire Precautions in Existing Places of Work that Require a Fire Certificate, Factories, Offices, Shops and Railway Premises* (HMSO, 1989). This guide replaces two earlier separate guides, one covering factories and the other offices, shops and railway premises. It sets out the basic standards for means of escape and other related fire precautions in existing premises which require a Fire Certificate.

There is also a suggested code of practice for buildings too small to require certificates: *Code of Practice for Fire Precautions in Factories, Offices, Shops and Railway Premises Not Required to Have a Fire Certificate* (HMSO, 1989). This gives practical guidance on standards for means of escape in the event of fire and means of fire-fighting in existing premises where the requirement for a Fire Certificate is waived, or there are insufficient numbers for a Fire Certificate to be necessary.

There is also a simple occupiers' and owners' guide to the requirements of the Fire Precautions Act 1971, entitled *Fire Safety at Work* (HMSO, 1989). This guidance is intended for owners, occupiers and managers of workplaces and contains good general advice on fire precautions for building designers.

The fire authority must consult with the relevant local authority before requiring work to be done prior to the issue of a Fire Certificate. There is also a statutory bar on the fire authority which prevents them from making requirements where at the time of the erection of the building it complied with Building Regulations which imposed requirements as to the means of escape.

Fire Certificates specify:

1. Means of escape to be provided.
2. Fire-fighting and fire escape requirements to be provided.

3. Fire alarm system to be provided.
4. Limitations on any particular explosive or flammable materials which may be stored on site.

They may also cover:

1. Maintenance conditions.
2. Staff training.
3. Limits on the number of people permitted in the building.
4. General fire precautions.

The Fire Safety and Safety at Places of Sport Act 1987 amended the Fire Precautions Act 1971 and established a class of premises which can be exempted from the necessity of obtaining a certificate. However, it requires a building's owners/managers to provide an adequate means of escape and fire-fighting equipment.

Although hospitals and residential care premises were not designated under the Fire Precautions Act 1971, two draft guides were produced outlining the standards which would be expected if designation occurred. They are: *Draft Guide to Fire Precautions in Hospitals* (Home Office/Scottish Home and Health Department, 1982) and *Draft Guide to Fire Precautions in Existing Residential Care Premises* (Homes Office/Scottish Home and Health Department, 1983). They outline the standards which would have been required if these premises had been designated under the Fire Precautions Act 1971. They have no statutory force, but act as the authoritative guidance on means of escape and related fire precautions in existing health care premises.

A further document which has become widely accepted – although it is not legally enforceable – covers houses in multiple occupation; this applies both to hostels and houses converted into self-contained dwelling units and deals with fire resistance of surfaces, means of escape, warning systems, fire-fighting equipment and staff training: *Guide to Means of Escape and Related Fire Safety Measures in Certain Existing Houses in Multiple Occupation* (Home Office/Department of the Environment/Welsh Office, 1988).

Obviously there is a substantial degree of overlap between the role of the fire authority and that of the building control authority, especially in regard to new buildings. One of the recommendations of the Bickerdike Allen Report (section 9.1) was that the different government departments (Department of the Environment, Welsh Office and Home Office) should jointly produce a procedural guidance document to clarify for designers the division of responsibility between the statutory authorities, and the approvals process which the architect should follow; this extremely useful document applies to England and Wales: *Building Regulations and*

Fire Safety, Procedural Guidance (Department of the Environment/Home Office/Welsh Office, 1992).

Health and safety at work

The Health and Safety at Work Act 1974 places responsibilities on employers, employees and others who may be affected by their actions for health, safety and welfare. The Act makes employers responsible for providing adequate information and training about safety matters. More particularly, it makes the Health and Safety Inspectorate responsible for policing requirements relating to the storage of hazardous substances (e.g. highly flammable liquids) and certain fire safety risks in manufacturing processes.

The Act also amends and strengthens the Fire Precautions Act 1971, as it applies to all premises 'used as a place of work' and with very few exceptions. In addition, it applies to existing as well as new buildings; all premises to which it applies require a certificate to be issued by the fire authority.

Section 2 of the Act places a duty of care on the employer towards the employees in respect of safety. The employer must provide information, training and supervision and also maintain the place of work as safe, and provide means of access and egress that are safe. The employer is required to have a published safety policy and consult with safety representatives from the employees. Section 7 of the Act places a duty of care on employees for their own safety.

The Health and Safety Executive administer the Act, and are responsible for the issuing of certificates for 'special premises' under the Fire Certificates (Special Premises) Regulations 1976. These have very similar provisions and requirements to certificates issued under the Fire Precautions Act 1971.

Licensing laws

Under the Licensing Act 1964, fire authorities can object to the issue of a licence to sell or supply intoxicating liquor or to refuse to register a club which supplies alcoholic drink if the means of escape are not satisfactory or there is undue risk of fire in the premises.

Under the Gaming Act 1968, which covers casinos and bingo halls, fire authorities can object to the granting or renewal of a licence if the premises are unsuitable by reason of their layout, character and location, or if reasonable facilities to inspect the buildings have been refused.

In the case of both alcohol and gaming licences, application must be to the licensing court (licensing justices), who must consult the fire authority. The fire authority makes observations and the court makes a ruling which may include requirements to alter the premises. An applicant may appeal to the court within 14 days.

Under the Cinematograph Act 1985 and the Theatre Act 1968, local authorities are responsible for ensuring adequate fire safety in places of public entertainment before granting licences. The regulations cover means of escape, lighting and general fire precautions.

Although there is not prescriptive legislation as to the details of what can be required under the various Licensing Acts, a guide was published in 1990 which sets out suggested standards: Home Office and Scottish Home and Health Department, *Guide to Fire Precautions in Existing Places of Entertainment and Like Premises* (HMSO, 1990). This recommends standards for means of escape and other fire precautions in a wide range of premises used for public entertainment and recreation. It was drafted by a working party set up following the Stardust Disco fire in Dublin and is aimed mainly at Building Control Officers and Fire Prevention Officers. It aims to set acceptable standards and encourages consistency in enforcement.

Other legislation

Fire safety provisions are attached to a plethora of Acts often related to specific building types or specific activities; these include:

Animal boarding establishments	Housing Acts
Caravan sites	Mental Health Acts
Children Acts	Nurseries
Education Acts	Petroleum
Explosives	Pipe lines
Fireworks factories	Social Work Acts.

As well as the national legislation which has been mentioned, there is also a considerable body of local legislation concerned with fire safety. Many counties have powers invested in them in respect of fire safety for different building types. Some local Acts were introduced at the beginning of the century and many were transferred from the old authorities to their successors when local government re-organization occurred in 1974. The principal local Acts applying to counties (or metropolitan districts within the old metropolitan counties) which include fire safety provisions are as follows:

Avon	Cornwall
Berkshire	Cumbria
Cheshire	Derbyshire
Clwyd	Dyfed
Cleveland	East Sussex

Essex	Mid-Glamorgan
Greater Manchester	South Glamorgan
Hampshire	South Yorkshire
Humberside	Stafforshire
Isle of Wight	Surrey
Kent	Tyne and Wear
Lancashire	West Glamorgan
Leicestershire	West Midlands
Merseyside	West Yorkshire.

The following district councils also have local Acts which make significant regulations about fire safety:

Bournemouth	Poole
Hereford	Plymouth
Kingston upon Hull	Worcester.

9.3 NORTHERN IRELAND LEGISLATION

Northern Ireland has its own legislative system, both for new and existing buildings, and therefore has to be considered separately.

New buildings are subject to the Building Regulations, which were revised and re-issued during 1990. Sadly, the opportunity to adopt the same Approved Documents, then being prepared as part of the revision of the Building Regulations and Approved Documents in England and Wales, was not taken. The production of different regulations and Approved Documents for each part of the UK must be one of the most pointless and time-wasting of government activities. It also ensures that architects working in more than one area are faced with the requirement to be familiar with more than one set of documentation. All of this confusion strengthens the argument for design from first principles; the nature of building and of fire does not vary with national boundaries, and the approximations and crude estimates in one set of rules are not likely to be more accurate than those in another.

The Building Regulations (Northern Ireland) 1990 have two sections dealing with fire safety. Part E covers 'Structural Fire Precautions', and consists of prescriptive regulations which have to be precisely followed, a total of 46 pages (plus 20 pages of schedules) of detailed instructions in legal form and without diagrams; it is very similar to the old form of the 1976 Regulations for England and Wales, abandoned in 1985. Part EE covers 'Means of Escape in Case of Fire' and is much more akin to the functional format of the present Building Regulations for England and Wales, with just a page and a half asking for 'reasonable' means of escape and giving the relevant British Standards as being 'deemed to satisfy' provisions.

There are already plans for a further revision of these Building Regulations, and it is hoped that this time the opportunity will be taken to link them to the England and Wales Approved Document B, so that consistency can be achieved.

Existing buildings are covered by the Northern Ireland equivalent of the Fire Precautions Act 1971, namely the Fire Services (Northern Ireland) Order 1984. This sets out virtually identical conditions and standards to the Fire Precautions Act, but the list of designated premises is different. So far, designations under this order are:

1. Leisure premises (1985).
2. Hotels and boarding-houses (1985).
3. Factory, offices and shop premises (1986).
4. Betting, gaming and amusement premises (1987).

There are also regulations (1986) which outline the fire precautions to be taken in smaller factory, office and shop premises which do not require a Fire Certificate under the designation order.

9.4 BRITISH STANDARDS AND INTERNATIONAL STANDARDS

British Standards are published by the British Standards Institute (BSI), prefixed with the letters BS. International standards are published by the International Organization for Standardization, prefixed with the letters ISO. The British Standards Institution is based at 2 Park Street, London W1A 2BS (071 629 9000).

BS 476: Fire Tests on Building Materials and Structures.
Part 3: 1975, External fire exposure roof test.
This test assesses the ability of a roof structure to resist penetration by fire when the outer surface is exposed to radiation and flame, and the likely extent of surface ignition during penetration. Roof structures are classified according to the actual times recorded, for example, 'a P60' designation means that the sample of roof resisted penetration for at least 60 min. Roof classifications are prefixed by a designation showing whether it was tested as a sloping (S) or flat (F) construction.
Part 4: 1970, Non-combustibility test for materials.
This text assesses whether materials are non-combustible by ascertaining if samples will give off heat or flame when heated.
Part 5: 1979, Method of test for ignitability.
This test assesses whether a material will ignite when subjected to a flame for 10 s.
Part 6: 1989, Method of test for fire propagation for products.
This text assesses the contribution of combustible materials to fire growth when they are subjected to flame and radiant heat.

Part 7: 1987, Method of classification of the surface spread of flame of products.
This test assesses the spread of flame across flat materials (normally wall or ceiling linings) when they are subjected to flame and radiant heat. Materials are classified as follows:

Class	Max. flame spread at:	
	1.5 mins	10 mins
1	165 mm	165 mm
2	215 mm	455 mm
3	265 mm	710 mm
4	900 mm	900 mm

Part 10: 1983, Guide to the principles and application of fire testing.
A basic guide.
Part 11: 1982, Method for assessing the heat emission from building products.
This test is a development of the basic test for non-combustibility in Part 4, and it quantifies the level of heat given off by a material when it is heated.
Part 12: 1991, Method of test for ignitability of products by direct flame impingement.
This test uses a choice of seven flaming ignition sources for a variety of flame application times.
Part 13: 1987, Method of measuring the ignitability of products subjected to thermal irradiance.
This test is a development of the basic test for ignitability in Part 5, and it measures the ease with which materials ignite when subjected to thermal radiation in the presence of a pilot ignition source. It is identical to ISO 5657: 1986, Fire tests. Reaction to fire. Ignitability of building products.
Part 20: 1987, Method for determination of the fire resistance of elements of construction (general principles).
Part 21: 1987, Method for determination of the fire resistance of loadbearing elements of construction.
Part 22: 1987, Method for determination of the fire resistance of non-loadbearing elements of construction.
Part 23: 1987, Method for determination of the contribution of components to the fire resistance of a structure.
These tests assess the fire resistance of different elements of construction and they replace Part 8: 1972, Test methods and criteria for the fire resistance of elements of building construction. The length of time for which the building elements can satisfy the following criteria, under test conditions, is recorded:

- **loadbearing** capacity (supporting the test load without excessive deflection);

- **integrity** (resisting collapse, the formation of holes, and the development of flaming on the unexposed face);
- **insulation** (resisting an excessive rise in temperature on the unexposed face).

Test results are normally expressed for each of the criteria in minutes. Columns and beams have only to satisfy the loadbearing criteria, and glazed elements normally only the integrity criteria, while floors and walls have to satisfy all three criteria.

Part 24: 1987, Method for determination of the fire resistance of ventilation ducts.

This test assesses the ability of ductwork to resist the spread of fire from one compartment to another without the assistance of dampers. Results are given in minutes for each of the criteria of stability, integrity and insulation. It is identical to ISO 6944: 1985, Fire resistance tests. Ventilation ducts.

Section 31.1: 1983, Methods for measuring smoke penetration through doorsets and shutter assemblies. Method of measurement under ambient conditions.

Based on Part 1 (1981), the ambient temperature test, of ISO 5925: Fire tests. Evaluation of performance of smoke control door assemblies.

This test assesses the likely amount of smoke penetration through shut doorsets and shutter assemblies. Results are given in air leakage in cubic metres per hour.

Part 32: 1989, Guide to full-scale fire tests within buildings.

The British Standards Institution has also published the following documents on fire testing.

PD 6496: 1981, A comparison between the technical requirements of BS476 Part 8 (1972) and other relevant international standards and documents on fire resistance tests.

PD 6520: 1988, Guide to fire test methods for building materials and elements of construction.

Other ISO standards relevant to testing which the architect might encounter include the following:
ISO 834: 1975, Fire resistance tests. Elements of building construction.
ISO 1182: 1983, Fire tests. Building materials. Non-combustibility tests.
ISO 1716: 1973, Building materials. Determination of calorific potential.
ISO 3008: 1976, Fire resistance tests. Door and shutter assemblies.
ISO 3009: 1976, Fire resistance tests. Glazed elements.
ISO 3261: 1975, Fire tests. Vocabulary.

BS 750: 1984, Specification for Underground Fire Hydrants and Surface Box Framers and Covers.

British Standards and international standards

BS 1635: 1990, Graphical Symbols and Abbreviations for Fire Protection Drawings.
BS 3169: 1986, Specification for First-Aid Reel Hoses for Fire-Fighting Purposes.
BS 4422: Glossary of Terms associated with Fire.
Part 1: 1987, General terms and the phenomena of fire.
Part 2: 1990, Building materials and structures.
Part 3: 1990, Fire detection and alarm.
Part 4: 1975, Fire protection equipment.
Part 5: 1989, Smoke control.
Part 6: 1988, Evacuation and means of escape.
Part 7: 1988, Explosion detection and suppression means.
Part 8: Terms specific to fire-fighting rescue services and handling hazardous materials (forthcoming).
Part 9: 1990, Terms associated with marine fires.
This BS is gradually being revised so as to conform with ISO 8421: Fire protection. Vocabulary.

BS 4547: 1972, Classification of Fires.
BS 4569: 1983, Method of Test for ignitability (surface flash) of pile fabrics and assemblies having pile on the surface.
BS 4790: 1987, Method of Determination of the Effects of a Small Ignition Source on Textile Floor Coverings (hot metal nut method).
BS 5041: Fire Hydrant Systems Equipment.
Part 1: 1987, Specification for landing valves for wet risers.
Part 2: 1987, Specification for landing valves for dry risers.
Part 3: 1975, Specification for inlet breechings for dry riser inlets.
Part 4: 1975, Specification for boxes for landing valves for dry risers.
Part 5: 1974, Specification for boxes for foam inlets and dry riser inlets.
BS 5266: Emergency Lighting
Part 1: 1988, Code of practice for the emergency lighting of premises other than cinemas and certain other premises used for entertainment.
BS 5268: Code of Practice for the Structural Use of Timber.
Section 4.1: 1978, Method of calculating fire resistance of timber members.
Section 4.2: 1989, Methods of calculating fire resistance of timber stud walls and joisted floor constructions.
BS 5274: 1985, Specification for Fire Hose Reels (Water) for Fixed Installations.
BS 5306 Fire Extinguishing Installations and Equipment on Premises.
Part 0: 1986, Guide for the selection of installed systems and other fire equipment.
Part 1: 1976, Hydrant systems, hose reels and foam inlets.
Part 2: 1990, Sprinkler systems.

Part 3: 1985, Code of practice for selection, installation and maintenance of portable fire extinguishers.
Part 4: 1986, Specification for carbon dioxide systems.
Section 5.1: 1992, Halon 1301 total flooding systems.
Section 5.2: 1984, Halon 1211 total flooding systems.
Section 6.1: 1988, Specification for low expansion foam systems.
Section 6.2: 1989, Specification for medium and high expansion foam systems.
Part 7: 1988, Specification for powder systems.
BS 5378: Safety Signs and Colours.
Part 1: 1980, Specification for colour and design.
Part 2: 1980, Specification for colorimetric and photometric properties of materials.
Part 3: 1982, Additional specifications.
BS 5395: Stairs, Lobbies and Walkways.
Part 1: 1977, Code of practice for the design of straight stairs.
Part 2: 1984, Code of practice for design of helical and spiral stairs.
Part 3: 1985, Code of practice for the design of industrial-type stairs, permanent ladders and walkways.
BS 5423: 1987, Specification for Portable Fire Extinguishers.
BS 5445: Components of Automatic Fire Detection Systems.
Part 1: 1977, Introduction.
Part 5: 1977, Heat sensitive detectors – point detectors containing a static element.
Part 7: 1984, Specification for point-type smoke detectors.
Part 8: 1984, Specification for high-temperature heat detectors.
Part 9: 1984, Methods of test of sensitivity to fire.
BS 5446: Specification for Components of Automatic Fire Alarm Systems for Residential Premises.
Part 1: 1990, Point-type smoke detectors.
BS 5499: Fire Safety Signs, Notices and Graphic Symbols.
Part 1: 1990, Specification for fire safety signs.
Part 2: 1986, Specification for self-luminous fire safety signs.
Part 3: 1990, Specification for internally illuminated fire safety signs.
(Note also ISO 6309: 1987, Fire protection. Safety signs.)
BS 5588: Code of Practice for Fire Precautions in the Design of Buildings.
Part 1: 1990, Code of practice for residential buildings.
Part 2: 1985, Code of practice for shops.
Part 3: 1983, Code of practice for office buildings.
Part 4: 1978, Code of practice for smoke control in protected routes using pressurization.
Part 5: 1991, Code of practice for fire-fighting stairways and lifts.
Part 6: 1991, Code of practice for places of assembly.
Part 7: Code of practice for atrium buildings (forthcoming).

Part 8: 1988, Code of practice for means of escape for disabled people.
Part 9: 1989, Code of practice for ventilation and air conditioning ductwork.
Part 10: 1991, Code of practice for shopping malls.
The British Standards Institute has also published the following documents to accompany BS 5588.
PD 6512: Use of Elements of Structural Fire Protection with Particular Reference to the Recommendations Given in BS 5588: Fire Precautions in the Design and Construction of Buildings.
Part 1: 1985, Guide to fire doors.
Part 3: 1987, Guide to the fire performance of glass.

BS 5720: 1979. Code of Practice for Mechanical Ventilation and Air Conditioning in Buildings.
BS 5725: Emergency Exit Devices.
Part 1: 1981, Specification for panic bolts and panic latches mechanically operated by a horizontal pushbar.
BS 5839: Fire Detection and Alarm Systems in Buildings.
Part 1: 1988, Code of practice for system design, installation and servicing.
Part 2: 1983, Specification for manual call points.
Part 3: 1988, Specification for automatic release mechanisms for certain fire protection equipment.
Part 4: 1988, Specification for control and indicating equipment.
Part 5: 1988, Specification for optical beam smoke detectors.
BS 5852: 1990, Fire Tests for Furniture.
Replaces BS 5852: Part 1 (1979) and BS 5852: Part 2 (1982) but these will still be referred to as they are cited in legislation.
BS 5950: Structural Use of Steelwork in Building.
Part 8: 1990, Code of practice for the fire protection of steelwork.
BS 6266: 1992, Code of Practice for Fire Protection for Electronic Data Processing Installations.
BS 6336: 1982, Guide to Development and Presentation of Fire Tests and their Use in Hazard Assessment.
BS 6387: 1983, Specification for Performance Requirements of Cables Required to Maintain Circuit Integrity under Fire Conditions.
BS 6459: Door Closers.
Part 1: 1984, Specification for mechanical performance of crank and rack and pinion overhead closers.
BS6535: Fire Extinguishing Media.
Part 1: 1990, Specification for carbon dioxide.
Section 2.1: 1990, Specification for halogenated hydrocarbons.
Section 2.2: 1989, Specification for safe handling of halogenated hydrocarbons.

Part 3: 1989, Specification for powder.
BS 6575: 1985, Specification for Fire Blankets.
BS 6651: 1992, Code of Practice for Protection of Structures against Lightning.
BS 7175: 1989, Specification for Resistance to Ignition of Bedcovers and Pillows by Smouldering and Flaming Ignition Sources.
BS 7176: 1991, Specification for Resistance to Ignition of Upholstered Furniture.
BS 7177: 1991, Specification for Resistance to Ignition of Mattresses, Divans and Bed Bases.
BS 8110: Structural Use of Concrete.
Part 1: 1985, Code of practice for design and construction.
Part 2: 1985, Code of practice for special circumstances.
Part 3: 1985, Design charts to accompany Part 1.
BS 8202: Coatings for Fire Protection of Building Elements.
Part 1: 1987, Code of practice for the selection and installation of sprayed mineral coatings.
Part 2: 1992, Code of practice for the use of intumescent coating systems.

9.5 GUIDANCE

The amount of guidance available to the design team is deceptively extensive. There are a large number of government organizations, trade associations and private companies offering information, guidance and advice on fire safety; however, this information is of variable quality and extremely patchy in its coverage. Some areas are very well covered (e.g. auto-suppression systems), while in other fields (e.g. smoke control) there is only a scattering of experts and information usable by the design team. One of the problems facing the architect is establishing what information is reliable and trustworthy, and where there is need to take further more specialist advice. This section cannot hope to cover all the guidance which is available, but it does seek to consider the main sources of information and to identify the most recent and useful publications from these different sources. Where the title of documents is not completely self-explanatory, an additional note has been added to guide designers as to the relevance and application of the materials.

The guidance has been listed under the organization producing it as this is the most likely way that designers will be able to trace what they need. It would have been possible to organize materials by reference to the fire safety tactic to which they relate, but this would have resulted in considerable duplication. The useful life of a code or guide is only about 10 years, and therefore there is a continually shifting body of information. However, as specific documents go out of date they are likely to be replaced by the organization concerned, another reason for structuring this section by organization. Addresses are also included, wherever possible, to enable practices to obtain their own copies of key documents.

Fire Research Station

Melrose Avenue
Borehamwood
Hertfordshire WD6 2BL
081-953 6177

The Fire Research Station (FRS) at Borehamwood is part of the Building Research Establishment and has been involved in fire safety research for over 40 years. There is a staff of some 150 who are involved both in fire research and the fire testing of buildings and products. There are facilities for full-scale fire tests, including an old airship hanger at Cardington which is capable of taking reconstructions of complete buildings.

The FRS is of value to architects both through its library and consultancy services (below) and because of its numerous publications. Many of these publications are research reports of limited value to architects in practice, but certain of the publications are useful reference documents and the designer may well find them being referred to by the statutory authorities when assessing designs. Some are full-scale books or reports, others shorter works (digests) or single-sheet information papers. The following list is far from comprehensive, but does identify the ones liable to be of some use in the design process.

Books, Reports and Digests

Smoke Control in Buildings: Design Principles (BRE Digest 260, 1982).
General guidance on the design of systems which will provide safe escape routes from buildings.
Studies of Human Behaviour in Fire: Empirical Results and their Implications for Education and Design (D. Canter, BR61, 1985).
Essentially a research report, but it does also include a model of human behaviour in fires. From this model some basic principles related to design and management are derived.
Fire Safety in Buildings (H. L. Malhotra, BR96, 1986).
Although this document is a report of a Department of Environment sponsored study to examine what technical changes should be made in the 1985 Building Regulations for England and Wales, its succinct and logical approach to the fundamental issues of fire safety design makes it useful reading for the design team. The author reviews the available strategies for achieving fire safety, summarizing the basic technical information. The issues of adequacy and equivalency are also cogently explained and a series of recommendations made for improvements in

the legislation. While not a design guide, this report does contain a wealth of useful information simply explained.

Fire Resistant Steel Structure: Free Standing Blockwork-filled Columns and Stanchions (BRE Digest 317, 1986).
Describes methods of providing 30 min of fire resistance to 'universal section' steel columns, with diagrams and tables.

Fire Doors (BRE 320, 1987).
Discusses the role played by fire doors, methods of assessing their performance and requirements under the then (1987) regulations and codes of practice, it deals only in general terms and does not give detailed guidance on construction.

Psychological Aspects of Informative Warning Systems (D. Canter et al., BR127, 1988).
Some general guidance useful for architects involved in the specification of communications systems.

Guidelines for the Construction of Fire-resisting Structural Elements (W. A. Morris et al., BR128, 1988).
Contains a set of tables of notional periods of fire resistance for structural elements based on current test data and information; it also includes a valuable summary of general fire safety information on the principal materials and elements of construction and the tables cover:

- masonry construction (solid, hollow and cavity walls);
- timber framed internal walls (loadbearing and non-loadbearing);
- timber or steel framed external walls (loadbearing and non-loadbearing);
- concrete columns, beams, and floors (both plain soffit and ribbed open soffit);
- encased steel columns and beams;
- timber floors.

Fire Performance of External Thermal Insulation for Walls of Multi-storey Buildings (B. F. W. Rogowski et al., BR135, 1988).
Description of experimental tests, leading to fundamental design recommendations on external thermal insulation. Tables of design recommendations for both sheeted and non-sheeted systems.

Aspects of Fire Precautions in Buildings (R. E. H. Read and W. A. Morris, 2nd Edn, BRE, 1988).
An extremely useful guide to certain aspects of passive fire precautions, particularly means of escape and structural fire protection. Unfortunately, it does not provide a comprehensive guide for designers; it has good sections on standard fire tests and also extensive coverage of the history of fire safety legislation.

Increasing the Fire resistance of Existing Timber Floors (BRE Digest 208, new edn 1988).

Guidance

Explains how periods of fire resistance of up to 1 h may be obtained by upgrading existing timber floors; it covers the addition of protection to the underside of the ceiling, over the floorboarding and between the joists.

Design Principles for Smoke Ventilation in Enclosed Shopping Centres (H. P. Morgan and J. P. Gardner, BR186, 1990).

The most useful document on smoke ventilation available for designers. It is an update on the 1979 FRS publication on the same subject and is the result of work carried out by FRS and Colt International Ltd. It is an invaluable explanation of basic smoke flow calculations and their application to large volume spaces, with easily understood drawings and not too many unnecessary calculations.

Experimental Programme to Investigate Informative Fire Warning Characteristics for Motivating Fast Evacuation (BR 172, 1990).

Research report on innovative forms of alarm systems, some limited design use to architects involved in major developments.

External Fire Spread: Building Separation and Boundary Distances (BR 187, 1991).

Describes the different methods for calculating adequate space separation between buildings. Prepared to support Approved Document B.

Fire Modelling (BRE Digest 367, 1991).

A short digest outlining a proposed fire model. Mixture of generalized statements and very precise calculations (of little value to designers).

Sprinkler Operation and the Effect of Venting: Studies Using a Zone Model (BR213, 1992).

Describes a mathematical model developed by FRS and Colt International Ltd to assess the interaction between sprinklers and smoke venting. Intended for the fire engineer rather than the architect.

Information Papers

Important Factors in Real Fires (IP20/84).
Fire Spread in Buildings (IP21/84).
Smoke Spread in Buildings (IP22/84).
Domestic Fire Deaths (IP23/84).
Incidence and Nature of Fires in Traditional and Framed Housing (IP9/85).
A Simplified Approach to Smoke Ventilation Calculations (IP19/85).
The Behaviour of People in Fires (IP20/85).
Ignition and Growth of a Fire in a Room (IP4/86).
Appraisal of Passive Fire Precautions in Large Panel System Flats and Maisonettes (IP 18/86).
Fire Safety Considerations in the Design of Structural Sandwich Panels (IP 4/87).
Assessing Life Hazard from Burning Sandwich Panels (IP18/87).

Selection of Sprinklers for high rack storage in warehouses (IP5/88).
Thermal bowing in fire and how it affects building design (IP21/88).
Assisted Means of Escape of Disabled People from Fires in Tall Buildings (IP16/91).
Statistical Studies of Fire (IP17/91).
The Development of a Fire Risk Assessment Model (IP 8/92).
False Alarms from Automatic Fire Detection Systems (IP 13/92).

Loss Prevention Council

The Loss Prevention Council (LPC) is involved in all aspects of loss prevention and control, including fire safety. It is funded by the Association of British Insurers and Lloyd's and has four component parts. The first is the LPC Technical Centre, which has comprehensive testing and fire testing facilities and is also involved in establishing new standards. The second part is the Loss Prevention Certification Board, which operates a number of certification schemes for products.

These first two parts of the LPC are based at Borehamwood on a site adjoining the FRS. They now include three well-known organizations: the Fire Offices' Committee (FOC), which issues standards for automatic sprinklers and fire protection equipment; the Fire Insurers' Research and Testing Organization (FIRTO), which tests fire protection equipment and building materials; and the Insurance Technical Bureau (ITB), which is a consultancy service.

The third component part of the LPC is the National Supervisory Council for Intruder Alarms, which is more concerned with security than with fire. The fourth and final component of the LPC is the Fire Protection Association (FPA), which provides information and advice on all aspects of fire safety, including fire safety design. It produces a number of valuable publications and an authoritative journal reviewing recent fires.

The advisory and consultancy aspects of these different bodies will be considered later, under this section the principal guidance materials available will be listed. The first general publication is *List of Approved Products and Services*, which is produced annually and provided free on request. Other publications are as follows:

LPC Rules (formerly published by the Fire Offices Committee) gives fundamental information related to loss prevention measures.

LPC Rules for Automatic Fire Detection and Alarm Installations for the Protection of Property (1991).
Design rules for fire alarm installations to protect property.

LPC Rules for Automatic Sprinkler Installations. To replace the 29th edition of the FOC rules for automatic sprinkler installations (1992).
This is the only document to include both the British Standard (BS 5306,

Part 2) and the insurers' additional requirements. Technical Bulletins to accompany the LPC Rules include:

- 'Sprinkler Systems for Dwelling Houses, Flat and Transportable Homes';
- 'Sprinkler Protection of Intensive Hanging Garment Stores';
- 'Automatic Sprinkler Pump Testing and Commissioning';
- 'Sprinkler Head Design Characteristics';
- 'Supplementary Requirements for Sprinkler Installations Operating in the Dry Mode';
- 'Automatic Sprinkler Protection to High-rise and Multiple-storey Buildings'.

LPC Code of Practice for the Construction of Buildings (1990).
This statement of insurers' rules for the fire protection of industrial and commercial buildings has been developed to compensate for what the insurance industry perceives as a relaxation in structural fire resistance required in the 1991 Approved Document B. It contains more onerous requirements in certain situations and has therefore been heavily criticized for introducing conflicting standards. What must be remembered is that the Building Regulations are concerned only with life safety, while the insurance industry is concerned about property protection.

LPC Recommendations (formerly published by the Fire Offices Committee) give practices which insurers expect to be followed during the building process and use of the building; these include:

- 'Recommendations for Emergency Power, Heating and Lighting' (1984);
- 'Fire Protection of Atrium Buildings' (1990);
- 'Loss Prevention in Electronic Data Processing and similar installations. Part 1, Fire Protection' (1990).

LPC Standards, Quality Schedules and Related Documents gives details of particular standards for different components and systems. These are mainly concerned with communication systems, extinguishing systems, fire doors, glazing and fire protection to steelwork.

The FPA has also published the following useful documents.
Heritage under Fire (with the UK Working Party on Fire Safety in Historic Buildings) (1990).
The Fire Protection of Old Buildings and Historic Town Centres (1992).
Fire Safety and Security on Construction Sites (1992).
Planning Company Fire Safety (1992).
The *Compendium of Fire Safety Data* (produced by the FPA) brings together advisory information on all aspects of fire, its prevention and control. The compendium consists of six volumes, and within each volume there are

separate data sheets on different subjects; these are revised and added on a regular basis:
1. 'Organization of Fire Safety: Management of Fire Risks';
2. 'Industrial and Process Fire Safety: Occupancy Fire Hazard';
3. 'Housekeeping and General Fire Precautions: Nature and Behaviour of Fire';
4. 'Information Sheets on Hazardous Materials';
5. 'Fire Protection Equipment and Systems';
6. 'Buildings and Fire'.

Obviously Volumes 5 and 6 are of most interest to the design team; however, any large architectural practice should have the complete compendium within its library.

Useful addresses:

Loss Prevention Council
140 Aldersgate Street
London EC1A 4HX
071-606 1050

LPC Technical Centre/Loss Prevention Certification Board
Melrose Avenue
Borehamwood
Hertfordshire WD6 2BJ
081-207 2345

Fire Protection Association
140 Aldersgate Street
London EC1A 4HX
071-606 3757

London District Surveyors' Association

Until the introduction of the 1985 Building Regulations, inner London had its own set of building by-laws administered by District Surveyors. The London District Surveyors' Association remains in existence and has produced two significant documents. The London Fire and Civil Defence Authority played a prominent role in the production of these two guides and they are therefore used as the basis for the appraisal of applications for approval as part of the formal consultation procedures. The two guides are:
Fire Safety Guide No 1: Fire Safety in Section 20 Buildings (1989).
This is based on a previous Greater London Council document, which has now been amended and rewritten to harmonize with the 1985 Building Regulations.
Fire Safety Guide No2: Fire Safety in Atrium Buildings (1989).

Department of Health, NHS Estates

The Department of Health, through NHS Estates, have produced a complete series of guidance documents on the fire safety of health care buildings. In legal terms, they are only advisory, but because of the move towards performance specification and Approved Documents in the Building Regulations, they have been recognized as representing adequate standards of fire safety and have become, effectively, 'law'. These documents are collectively known as FIRECODE, published by HMSO; the series is as follows:

Policies and Principles (1987).
This contains basic policy, principles and key management guidance. It is a short document intended for health service managers, as well as designers. Currently being revised.

Fire Precautions in New Hospitals (Health Technical Memorandum, HTM 81, 1987).
This sets out the fire safety requirements for all new hospitals, except those built under the NUCLEUS system.

NUCLEUS Fire Precautions Recommendations (1989).
This sets out the fire safety requirements for all new hospitals built under the NUCLEUS system.

Alarm and Detection Systems (HTM 82, 1989).
This provides specifications for alarm and detection systems both for new hospitals and the upgrading of existing hospitals.

Fire Safety in Health Care Premises (HTM 83, 1982).
This provides guidance for fire safety officers within health care buildings and so is of less relevance to designers. Currently being revised.

Assessing Fire Risk in Existing Hospital Wards (HTM 86, 1987).
This outlines a method of assessing the fire safety in existing premises in order that the most serious areas can be identified for upgrading. It is one of the very few systematic methods available to assess fire safety on a true fire engineering basis. Currently being revised.

Textiles and Furniture (HTM 87, 1989).
This provides guidance on the specification of furniture, fabrics and fittings for health care premises. Currently being revised.

Guide to Fire Precautions in NHS Housing in the Community for Mentally Handicapped and Mentally Ill People (HTM 88, 1986).
This provides guidance on fire precautions for new purpose-built buildings for the care of mentally handicapped and mentally ill patients within the community.

Laundries (Fire Practice Note, FPN 1, 1987).
Storage of Flammable Liquids (FPN 2, 1987).
Escape Bed Lifts (FPN 3, 1987).
Commercial Enterprises on Hospital Premises (FPN 5, 1992).

Directory of Fire Documents (1987).
This lists documents relating to fire in health care premises which were relevant in 1987.

Although not technically part of FIRECODE, two other documents are particularly relevant to the fire safety of health care premises and essential to the design team if they are engaged in the upgrading of existing premises. These are the two draft guides already mentioned under the Home Office section covering existing hospitals and residential care premises. They are being revised and will form HTM85.

Department for Education

The former Department of Education and Science has published within its building bulletins series an extremely good guide to fire safety in educational buildings, entitled *Fire and the Design of Educational Buildings* (Building Bulletin 7, HMSO, 1988). Building Bulletin 7 was first published in 1952, the current sixth edition has been updated to complement the Approved Documents issued with the 1985 Building Regulations for England and Wales. It has legal status and, in effect, provides the Code of Practice for fire safety in educational buildings. It is primarily aimed at new design work, but may also be appropriate in the context of alterations to existing buildings. Fortunately, it has an excellent 'first principles' approach and is clearly illustrated and explained; it covers:

- means of escape;
- precautions against fire;
- structural fire precautions;
- fire warning systems and fire-fighting;
- fire safety management issues within the buildings.

Timber Research and Development Association

Stocking Lane
Hughenden Valley
High Wycombe
Buckinghamshire HP14 4ND
0494 563091

The Timber Research and Development Association (TRADA) is a research and testing organization for the timber industry. It produces a series of valuable publications for architects, the Wood Information Sheets (WIS), and provides test certificates for certain products:

Flame Retardant Treatments for Timber (WIS 2/3-3, 1988, under rev. 1992).
Describes surface finishes and impregnation treatments for timber and wood based sheet materials, and relates these to the relevant British Standards.

Low Flame Spread Wood-based Board Products (WIS 2/3-7, 1987, under rev. 1992).
Describes products which have treatments applied to the particles veneers or fibres during or after manufacture such that the board has low surface spread of flame properties.
Timber and Wood Based Sheet Materials in Fire (WIS 4-11, rev. 1991).
Describes the behaviour of timber in fire.
Technology of Fire Resisting Doorsets (WIS 1-13, 1989).
Describes the performance requirements for fire-resisting doorsets, and gives guidance on door furniture and glazing.
Fire Resisting Doorsets by Upgrading (WIS, 1-32, rev. 1991).
Gives guidance on the assessment of existing doorsets for upgrading and on upgrading techniques.
Timber Building Elements of Proven Fire Resistance (WIS 1-11).
A series of sheets giving details of timber constructions which have had their fire resistance validated by test:

- D5 to D12: doors (rev. 1991);
- M1 to M4: glazed screens (1983 and 1987).

The Steel Construction Institute

Silwood Park
Ascot
Berkshire SL5 7QN
0344 23345

The Steel Construction Institute (SCI) is a research and development organization for the steel construction industry. It has produced a number of useful and detailed documents on the fire protection of steelwork, as well as a couple of technical reports giving test data and an excellent fire report.

Fire Protection for Structural Steel in Buildings (2nd edn, 1992).
Compilation of techniques of fire protection, published by the Association of Structural Fire Protection Contractors and Manufacturers (ASFPCM), with the Steel Construction Institute and Fire Test Study Group.

The Fire Resistance of Composite Floors with Steel Decking (G. M. Newman, 1989).
Outlines a fire engineering method of verifying fire resistance.

Fire and Steel Construction: The Behaviour of Steel Portal Frames in Boundary Conditions (G. M. Newman (2nd edn, 1990).
Describes the behaviour of portal frames in fire, and gives guidance on the design of column bases to resist rafter collapse.

Fire Resistant Design of Steel Structures – a Handbook to BS 5950 Part 8 (R. M. Lawson and G. M. Newman, 1990).

A user-friendly guide to the British Standard with examples of the evaluation of fire resistance.
Technical Report: Enhancement of Fire Resistance of Beams by Beam to Column Connections (R. M. Lawson, 1990).
Technical Report: Fire Resistance of Composite Beams (G. M. Newman and R. M. Lawson, 1991).
Investigation of Broadgate Phase 8 Fire (SCIF, 1991).

Textbooks

There are surprisingly few textbooks on the fire safety of buildings and anything published more than 15 years ago will already be out of date, both in terms of legislation and building types and materials. The textbooks listed below are those which may be of some value to the architect and the design team, although only certain parts of each are relevant.
Smoke Control in Fire Safety Design (G. Butcher and A. Parnell, E. and F. N. Spon, London, 1979).
Fairly technical work on smoke control. It provides a lot of useful information, but it is not an exhaustive design guide.
Design of Fire Resisting Structures (H. L. Malhotra, Surrey University Press, Guildford, 1982).
Too technical to be of value to architects, but perhaps useful for engineers. An attempt is made to explain simply the details of structural design, but it still probably uses too many symbols and formulae.
Atrium Buildings (G. Butcher *et al.*, Architectural Press, London, 1983).
Extremely good chapter on fire safety in atrium buildings ('Design for fire safety'). Possibly one of the best introductions to this very complex subject.
Designing for Fire Safety (G. Butcher and A. Parnell, Wiley, Chichester, 1983).
This is an excellent book which attempts to provide both information on fire science and to provide detailed guidance for architects, indicating what is relevant at each stage of the design process. Unfortunately, the scientific sections are possibly too detailed to be of value in design, and the detailed guidance is now dated because of changes in legislation and British Standards.
Fire and Building (AQUA Group, Granada, London, 1984).
A fairly good general textbook. The emphasis is on insurance and management aspects, rather than on fire safety design. It includes a useful glossary.
Buildings and Fire (T. J. Shields and G. W. Silcock, Longman Scientific and Technical, London, 1989).

Consultancy and advisory services

A very technical book most suitable for the specialist fire safety engineer, but probably too detailed for the average architect.
Underdown's Practical Fire Precautions (3rd edn, rev. R. Hirst, Gower, Aldershot, 1989).
A reference handbook aimed at fire engineers rather than designers. Particularly useful for the sections on the technology of active fire-fighting systems.
Fire Safety and the Law (J. Holyoak, A. Everton and D. Allen, 2nd edn, Paramount, Borehamwood, 1990).
As the subtitle describes it, this is a 'guide to the legal principles related to the hazard of fire'. This is a legal rather than technical book, but the chapters on current legislation and professional liabilities would be of interest to designers.
Fire from First Principles (Paul Stollard and John Abrahams, E. and F. N. Spon, London, 1991).
Designed as an undergraduate textbook for students of architecture and building, this provides a straightforward introduction to fire safety design. It includes a series of rules-of-thumb on escape and containment issues to aid student projects, and a clear introduction to fire science and fire engineering.
Manuals of Firemanship (Home Office, London).

'Book 1: Elements of Combustion and Extinction' (1974);
'Book 8: Building Construction and Structural Fire Protection' (1975);
'Book 9: Fire Protection of Buildings' (2nd edn, 1990).

These are all very basic, but as they form the official textbooks for Fire Prevention Officers within the Fire Brigades, it is perhaps necessary for architects and designers to be aware of their existence.
Knight's Guide to Fire Safety Legislation (Charles Knight, London).
This is not a textbook, but a useful collection of all relevant legislation in a loose-leaf format. It was first issued in 1990 and will have to be regularly updated if it is to remain useful.
The Construction Industry Information and Research Association (CIRIA) publications on the fire resistance of concrete are research reports; however, with a shortage of good guidance on the use of concrete, these may be useful to designers:
Spalling of Concrete in Fires (H. L. Malhotra, TN118, 1984).
Fire Resistance of Ribbed Concrete Floors (R. M. Lawson, R107, 1985).
Fire Resistance of Composite Slabs with Steel Decking – Data Sheet (SP42, 1986).
Fire Tests on Ribbed Concrete Slabs (TN131, 1987).

9.6 CONSULTANCY AND ADVISORY SERVICES

The previous sections have all dealt with written guidance materials, but it may be necessary for the design team to seek direct help from a fire safety consultancy. In this case, there are five options for the design team to consider: the FRS, the LPC, a commercial testing station, a private fire safety consultancy or a trade association.

The Fire Research Station offers two levels of consultancy. At its simplest, their library services will assist architects and designers to find the correct documents and guidance from among their own and any other publications. They also offer a technical consultancy which will advise directly on the fire engineering of a particular design. This might range from the complete integrated smoke control system to the likely performance of a single component or assembly. They have an expert staff and excellent test facilities for experimental work.

The constituent parts of the Loss Prevention Council offer different services. The FPA has an information service which provides free advice and guidance to member organizations of the FPA; the LPC Technical Centre offer full facilities for the fire testing of building and assemblies; and the Loss Prevention Certification Board will give information on products which have been certified under the schemes.

In addition to the FRS and LPC, there are three main commercial testing stations to which architects can turn for fire testing and certification, all of which are registered with the National Measurement Accreditation Service. The first of these is SGS Yarsley Ltd (formerly Yarsley Technical Centre) in Surrey; the second is the fire research and testing laboratory of British Gypsum; and the third is the Warrington Fire Research Centre, which is the largest independent fire consultancy, fire research and fire testing organization in the U.K.

SGS Yarsley Ltd
Trowers Way
Redhill
Surrey RH1 2JN
0737 765070

British Gypsum Fire Research and Testing Laboratory
Eastleake
Loughborough
Leicestershire LE12 6JT
0602 214161

Warrington Fire Research Centre Ltd
Holmesfield Road
Warrington
Cheshire WA1 2DS
0925 55116

Most architects who seek fire safety advice will approach one of the

Consultancy and advisory services

independent fire safety consultancies now offering this service. As with consultants, some are excellent and some are little more than charlatans. It is essential for architects to decide what service they need and therefore the right consultancy to provide such services; it may be that the architects are only seeking someone to check their own designs for legislative compliance, but they may need to use consultants who are capable of the much more demanding task of preparing a full fire engineering solution to an unusual design problem. One useful guide to the architect in seeking the right consultant is to establish which professional organizations the consultants are members of.

The longest-established professional organization is the Institution of Fire Engineers (IFE) but most members tend to be Fire Brigade officers, not necessarily experts on fire safety design. Membership is by examination, but the standard of the corporate membership exam does not qualify candidates for Chartered Engineer status:

Institution of Fire Engineers
148 New Walk
Leicester LE1 7QB
0533 553654

The Institute of Fire Safety (IFS) was formed in 1992 by the merger of the Society of Fire Protection Engineers (SFPE) and the Society of Fire Safety Engineers (SFSE). The SFPE began in the USA, but in the 1980s established a number of British branches serving mainly engineers involved in designing fire protection systems. The SFSE was established in the late 1980s to attract fire safety designers, academics and engineers, and with its strict selection procedure it has hoped to achieve rapidly Chartered Engineer status. The merger of these two bodies to form the Institute of Fire Safety has created one professional body encompassing almost all consulting fire engineers in the U.K. and it is likely to achieve Chartered Engineer status within a short time.

Institute of Fire Safety
PO Box 687
Croydon CR9 5DD
081 654 2582

Although trade associations are established to serve the interests of their members and therefore may not always offer completely independent advice, they can provide a useful source of information to designers. This is particularly true when it comes to the selection of materials or components to meet an already determined specification. Trade associations in the field of fire safety which architects might wish to consult are as follows:

Association of Structural Fire Protection Contractors and Manufacturers (ASFPCM)
P.O. Box 111
Aldershot
Hampshire GU11 1YW
0252 21322

British Fire Protection Systems Association (BFPSA)
48a Eden Street
Kingston-upon-Thames
Surrey KT1 1EE
081 549 5855

Fire Extinguishing Trades Association (FETA)
48a Eden Street
Kingston-upon-Thames
Surrey KT1 1EE
081 549 8839

Fire Resistant Glass and Glazed Systems Association
20 Park Street
Princes Risborough
Buckinghamshire HP27 9AH
0844 275500

Intumescent Fire Seals Association (IFSA)
20 Park Street
Princes Risborough
Buckinghamshire HP27 9AH
0844 275500

Lastly it is important to include CERTIFIRE, this is the Certification Authority for passive fire protection products and services.

CERTIFIRE
101 Marshgate Lane
London E15 2NQ
081 555 3234

Appendix: Details of the 'Design Against Fire' course

The original series of 'Design against Fire' seminars on which this book is based was run in the Michaelmas term 1991 at the Queen's University of Belfast. It was designed by an *ad hoc* grouping of the various relevant groups (Fire Brigade, building control and local architects), who worked together to establish a joint continuing professional development course. The course was unique in being both completely multi-disciplinary and recognized by a university postgraduate certificate.

The course was not intended for inexperienced students, but rather experienced job architects, divisional Fire Prevention Officers and senior Building Control Officers. The course members worked together and learnt not only directly from the lectures, projects and visits, but most importantly from each other. The eight one-day seminars were held once a week and each seminar concentrated on one of the key fire safety strategies – e.g. containment, communications and escape.

The first session attempted to broaden the horizons of course participants by looking at fire safety as a science and as an aspect of risk engineering. All too often, courses on fire safety begin with regulations rather than reasons. In fire engineering it is the reasons and the principles which are most important. There were lectures from Jack Anderson, Dougal Drysdale and Paul Stollard, and these are to be found in the Introduction, and Chapters 1 and 2 of this book.

There was also the first of the group projects, and to maintain the first principles approach, this involved a risk analysis of an existing university building for which full plans were provided. The course members worked in syndicates of four, each with two designers, one Fire Prevention Officer and one Building Control Officer. They were not expected to carry out a full 'inspection' or produce detailed proposals for improvements, rather what was required was an assessment of possible problem areas and possible solutions. The syndicates were asked to work through the five principal fire safety tactics, thereby building up a fire safety strategy for the building. At the end of the workshop session, syndicates

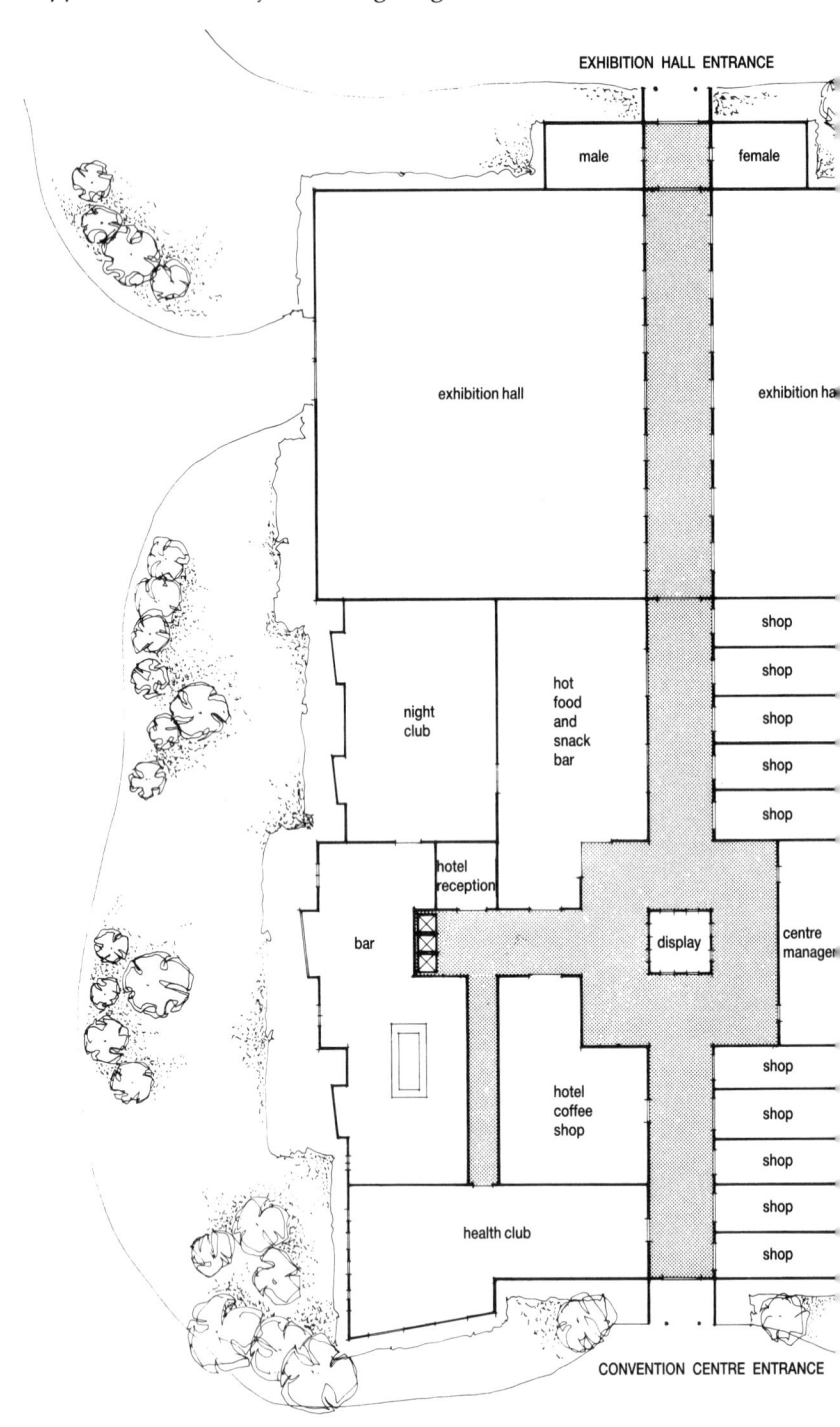

Appendix: Details of the 'Design Against Fire' course

EXHIBITION HALL ENTRANCE

male | female

exhibition hall

exhibition hall

shop
shop
shop
shop
shop

night club

hot food and snack bar

hotel reception

bar

display

centre manager

shop
shop
shop

hotel coffee shop

health club

shop
shop

CONVENTION CENTRE ENTRANCE

LAGANSIDE CONVENTION AND EXHIBITION CENTRE

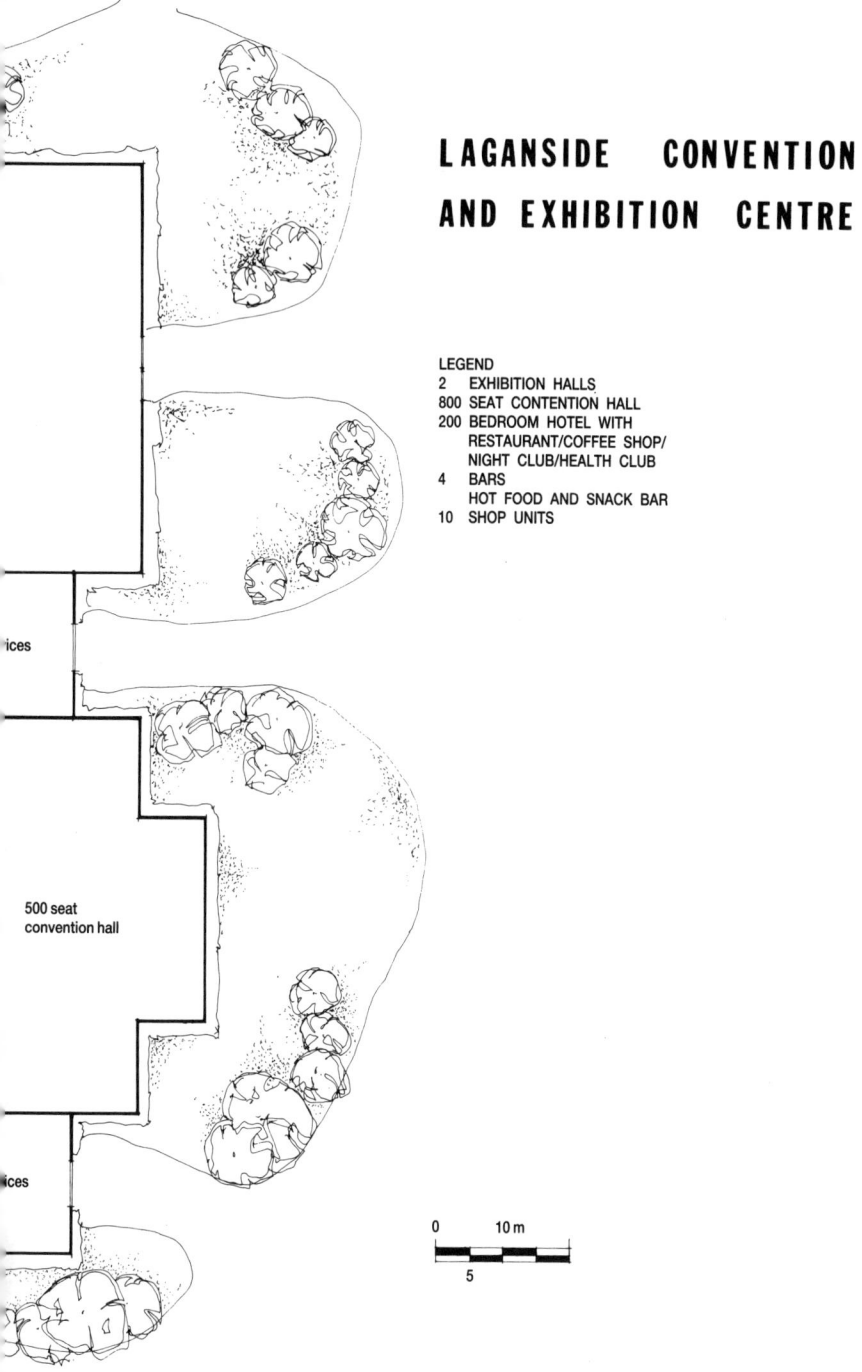

LEGEND
2 EXHIBITION HALLS
800 SEAT CONTENTION HALL
200 BEDROOM HOTEL WITH RESTAURANT/COFFEE SHOP/ NIGHT CLUB/HEALTH CLUB
4 BARS
   HOT FOOD AND SNACK BAR
10 SHOP UNITS

500 seat convention hall

0   5   10 m

Figure A1 Laganside Convention and Exhibition Centre: ground-floor plan.

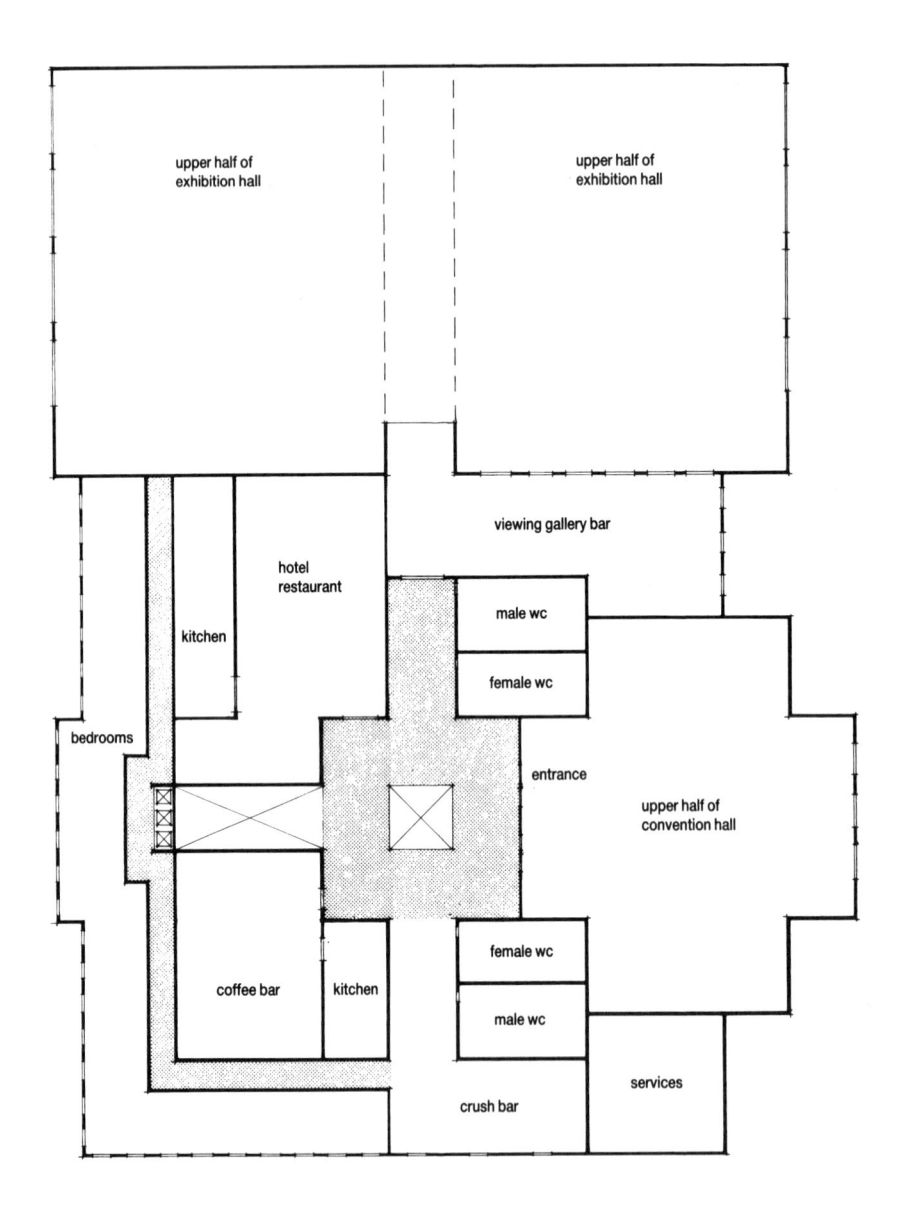

156 *Appendix: Details of the 'Design Against Fire' course*

- upper half of exhibition hall
- upper half of exhibition hall
- viewing gallery bar
- hotel restaurant
- kitchen
- male wc
- female wc
- bedrooms
- entrance
- upper half of convention hall
- coffee bar
- kitchen
- female wc
- male wc
- crush bar
- services

FIRST FLOOR

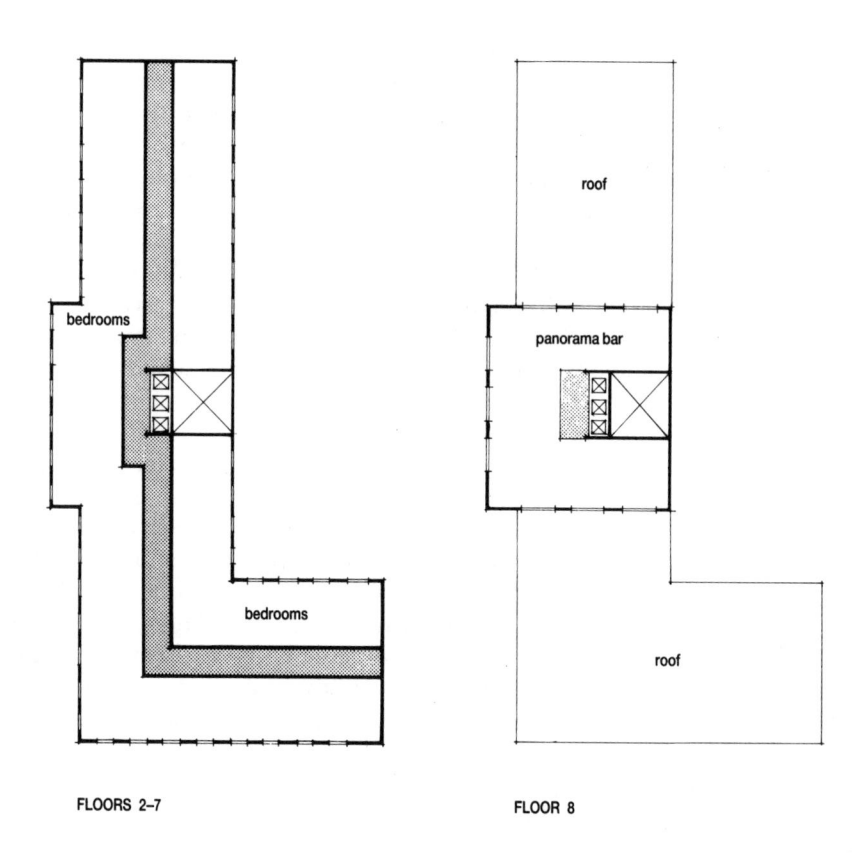

Appendix: Details of the 'Design Against Fire' course 157

FLOORS 2–7

FLOOR 8

Figure A2 Laganside Convention and Exhibition Centre: upper-floor plan.

compared fire safety assessments and strategies and it was interesting to see that the three professional groups were already beginning to drop their traditional confrontational roles.

When it came to fire-fighting and fire extinguishment, in session 2, those on the course were given the opportunity to deal with a real fire at the Fire Brigade's training school. Having experienced together the difficulties of fighting even a training fire, in thick smoke and wearing breathing apparatus, the course members found it somewhat easier to work together on the design projects.

With three distinct groups on the course, it was always intended that some parts would be familiar territory for each group. So the fire officers were of course very used to this situation and were able to act as tutors in this session, while in the lectures on detailed legislation building control were expected to take the lead and, in the design projects, it was hoped the architects would be acting as the tutors.

The third session addressed human behaviour in fire and fire escape. The keynote lecture was from Jonathan Sime (Chapter 5), and this was supported by lectures from local building control and Fire Brigade officers on the specific Northern Ireland legislation regarding means of escape. This combination of keynote lectures on principles complemented by more detailed lectures on the local situation was to be a feature of all the remaining sessions.

This session also included the first of three site visits, in this instance to a virtually complete shopping centre/leisure complex where escape issues were very important. Each of the site visits presented real fire engineering problems and it was possible to discuss these with the actual building control officers and fire brigade officers involved presenting their decisions and reasoning.

For the fourth session, the course moved to Londonderry. It was important that different venues should be used, so that there was some fairness in the distance that each course member had to travel, and also to permit case studies to be taken from different cities. The subject in this session was the integration of escape and containment issues and the keynote lecture was from John Abrahams (Chapter 6). There was also a series of case studies presented by Margaret Law, of Arups, on the integration of fire engineering solutions into the general design process. The site visit was to a converted mill building, which had a variety of serious problems.

There was a syndicate project on the design of means of escape for a conference/exhibition building (Figures A1 and A2), each syndicate being asked to identify the likely user groups, and then for each of the groups to estimate alertness, numbers, mobility, probable familiarity with the building and likely response to a fire alarm. Once the user groups had been analysed, the syndicates were to consider the appropriate escape

Appendix: Details of the 'Design Against Fire' course 159

strategy for each group, either egress or refuge, and the acceptable times for achieving safety. Only with the strategy established were the syndicates allowed to make more detailed proposals relating to horizontal and vertical escape routes, refuge areas, escape lighting and signing.

The fifth session moved back to Belfast and concentrated on passive containment with a keynote lecture by Bill Malhotra (Chapter 7). The project used the same building as in the previous week, but this time the groups were required to identify the likely fuel load in each area (and therefore the maximum compartment size), areas of high ignition risk which require additional compartmentation and particular user groups who will require special protection.

As the next session concentrated on communications, it was held at the Fire Brigade Headquarters in Lisburn. There was a visit to the brigade control room, a keynote address by John Northey (Chapter 4) and technical presentations by one of the large manufacturers and designers of fire detection and alarm systems.

The seventh day was devoted to smoke control systems and began with a keynote address by Howard Morgan (Chapter 8). There was a visit to a recently completed shopping centre in the middle of Belfast with talks from the architects, managers, fire officers and building control officers involved.

The project was a little different as this time the syndicates were asked to consider how the present system of fire safety design, regulation and approval could be improved in Northern Ireland. Just as with each of the technical projects where they were asked to think from first principles, so in this last group project they were required to take a radical look at the management of fire safety. The proposals from each syndicate had to have the unanimous support of all members and therefore be cross-disciplined. The agreed set of proposals from all the syndicates was widely circulated in Northern Ireland to act as a discussion document; it consisted of the following suggestions:

- There is a need for the setting up of a project team on large or abnormal projects (e.g. mixed-use developments). Such a project team must include: the architects, Building Control Officers and Fire Prevention Officers. It should first meet during the sketch design stage, and then at formally agreed times for the rest of the project.
- Consideration should be given to the feasibility of locating the Fire Prevention Officers for each division and the Building Control Officers for the corresponding authorities in the same building.
- Joint training of Building Control Officers and Fire Prevention Officers should be increased, both at the recruit and at higher levels.
- There should be discussions on the possibility of Building Control

Officers and Fire Prevention Officers developing areas of complimentary specialisms.
- The existing Part E committee should be strengthened by the introduction of additional members, so that it can become the authoritative fire safety committee for Northern Ireland. It should also be enabled to disseminate its advice and interpretations more widely to architects, as well as the statutory authorities.
- On all buildings there should be contact between the architects and statutory authorities at the earliest practical stage.
- There should be the possibility of stage approval for 'fast-track' projects. This would require a rigid and enforceable timescale to which all worked.
- Each completed building should be provided with a Management Manual, detailing how fire safety within the building is intended to be managed, and outlining the maintenance requirements of the fire safety systems.
- Consideration should be given to 'tidying up' anomalies in geographical boundaries, so that the fire prevention divisions and building control groups are contiguous.
- Changes in legislation should be such as to promote a common 'rulebook' for fire regulations, so that the identical guidance is used in Northern Ireland as in England and Wales, and in Scotland.
- A Guidance Procedure document should be produced which sets out clearly and simply the procedures for achieving statutory approval for the fire safety of different building types, and explaining the inter-relationship of the different statutory authorities.
- Fire safety awareness courses for building owners, managers and developers should be encouraged and promoted.

The final session consisted of two distinct parts. In the morning there were a number of lectures on issues closely related to fire engineering, including arson (Chapter 3), insurance, and health and safety. There was also a lecture outlining sources of advice (Chapter 9) and summaries from building control and the Fire Brigade. In the afternoon each course participant was required to make a five-minute presentation, or course response. This was compulsory as it was the formal assessed part for the university certificate. Course members had to prepare a short contribution based on what they had gained from the course, or on how fire safety design was relevant to a current project on which they were working.

The course was considered a success by the participants and one well worth repeating. On the basis of their suggestions, changes were able to be made in the detailed planning of the future courses, but the basic principles remain unaltered.

Glossary

acceptability: the extent to which the fire safety meets that deemed by society to be necessary; this may be expressed through legislation and influenced by the frequency of disasters.

active containment: measures to contain the spread of fire which require the operation of some form of mechanical device – e.g. the operation of smoke vents or the release of fire shutters.

addressable system: a comprehension/analysis system which compares the present situation with data stored in the system's memory, to derive more than just fault and fire signals from the detectors and to permit the precise location of the fire to be established.

alternating sprinkler system: a sprinkler system which can be changed from 'wet' operation in summer to 'dry' operation in winter.

auto-suppression: fire extinguishing systems which activate automatically on detecting a fire.

bridgeheads: intended safe bases for fire-fighters attempting to tackle a fire within the building, served by protected lifts and adjacent to rising mains.

burning brands: flaming debris carried by convection currents from buildings on fire.

combustibility: the ease with which a material will burn when subjected to heat from an already existing fire.

combustion: the series of rapid chemical reactions between a fuel and oxygen releasing heat and light.

communication: the fire safety tactic of ensuring that if ignition occurs the occupants are informed immediately and any active fire systems are triggered.

compartmentation: the technique of dividing a building into a number of compartments or sub-compartments.

compartments: fire and smoke tight areas into which a building can be divided to contain fire growth and limit travel distance, offering 60 min of resistance.

conduction: the transfer of heat by direct physical contact between solids.

containment: the fire safety tactic of ensuring that the fire is contained to the smallest possible area, limiting the amount of property likely to be damaged and the threat to life safety.

convection: the transfer of heat by the movement of the medium in liquids and gases.

conventional system: a comprehension/analysis system where it is only possible to establish fault or fire signals from the detectors.

diffusion flames: flames from combustion, where the rate of burning is determined by the rate of mixing of flammable vapours from a solid or liquid fuel and oxygen.

dry rising main: a rising main normally kept empty of water, but to which the fire service can supply water at ground level in the event of a fire.

dry sprinkler system: a sprinkler system where the majority of pipework is air filled until it is triggered.

emergency lighting: lighting provided by standby generators on the failure of the mains supply.

envelope protection: the limitation of the threat posed by a fire to adjoining properties and to people outside the building, and the threat posed by a fire in an adjoining property.

equivalency: the provision of the same level of fire safety by different combinations of fire safety measures.

escape: the fire safety tactic of ensuring that the occupants of the building and the surrounding areas are able to move to places of safety before they are threatened by heat and smoke.

escape lighting: lighting provided on the failure of the normal lighting circuits by self-contained fittings.

extinguishment: the fire safety tactic of ensuring that the fire can be extinguished quickly and with minimum consequential damage to the building.

fire: the combustion process accompanied by the production of smoke and/or flame.

fire engineering (or holistic fire safety design): design which considers the building as a complex system, and fire safety as one of the many interrelated sub-systems which can be achieved through a variety of equivalent strategies.

fire growth curve: the relationship between the time from ignition and the size of the fire.

fire load: the amount of fuel within a room or building which will burn to release heat and feed the growth of the fire.

fire point: the temperature to which a fuel has to be heated for the vapours given off to sustain ignition if an ignition source is applied.

fire propagation: the degree to which a material will contribute to the spread of a fire through heat release, when it is itself heated.

fire protection: measures to limit the effects of fire, including the fire

safety tactics of communication, escape, containment and extinguishment.
fire resistance: the capability of a component or an assembly of components to resist fire by retaining their loadbearing capacity, integrity and insulating properties.
fire risk: the probability of ignition, and of consequent life and property loss.
fire safety components: the specific building elements, structures and procedures, which the architect can use tactically to achieve fire safety – e.g. they may include fire doors, sprinklers, escape stairs, and fire drills.
fire safety objectives: the objectives which the architect must satisfy in order to achieve a fire-safe building, normally life safety and property protection.
fire safety tactics: the tactics which the architect can adopt in order to satisfy the fire safety objectives, normally prevention, communications, escape, containment and extinguishment.
flame: the visible manifestation of the reaction between a gaseous fuel and oxygen.
flammable limits: the range of concentrations of flammable vapours to air within which a flame will be produced in the presence of an ignition source.
flashover: the decisive moment in the growth of a fire within a compartment when all combustible surface materials rapidly reach their fire points and ignite within 3–4 s.
flash point: the temperature to which a fuel has to be heated for the vapours given off to flash if an ignition source is applied.
fuel limitation: the control of the amount of fuel within a building or room.
fuel load: the amount of potential fuel within a building or room; this includes both the building's fabric and contents.
holistic fire safety design (or fire engineering): design which considers the building as a complex system, and fire safety as one of the many interrelated sub-systems which can be achieved through a variety of equivalent strategies.
ignitability: the ease with which a material can be ignited when subjected to a flame.
ignition: the process whereby oxygen and a gaseous fuel react under the influence of heat to initiate combustion.
ignition prevention: measures taken to reduce the probability of ignition.
ignition risk: the probability of ignition.
ignition source: a heat source, or flame, which will cause ignition.
ionization detector: a smoke detector which can identify the reduction

in electrical current across an air gap in the presence of a small radioactive source when smoke particles are present.

insulation: the resistance offered by a material to the transfer of heat.

integrity: the ability of a material to resist thermal shock and cracking, and to retain its adhesion and cohesion.

intumescents: materials which react to heat by expanding and forming an insulating layer.

laminated glass: glass which incorporates layers of transparent and translucent intumescent material.

loadbearing capacity: the dimensional stability of a material.

manual call point (or break glass call point): an alarm switch which can be activated by the occupants.

neutral plane: the level within a building where the internal air pressure is equivalent to atmospheric pressure.

occupancy load factor: a measure for calculating the likely number of people in particular building types for a given floor area.

optical detector: a smoke detector which can identify the reduction in light received from a light source by a photoelectric cell when smoke particles are present.

passive containment: measures to contain the spread of fire which are always present and do not require the operation of any form of mechanical device – e.g. the fire protection provided to the building structure or the fire-resisting walls provided to divide a building into different compartments.

phased evacuation: the planned evacuation of a building in stages.

place of safety: an area to which the occupants can move, where they are in no danger from fire.

pre-action sprinkler systems: one which is 'dry', but where water is allowed in on a signal from a more responsive detector (usually smoke) in advance of the heads being triggered.

pre-mixed flames: flames from combustion where the fuel is a gas and is already mixed with oxygen.

pressurization: the technique whereby staircases or corridors are pressurized such that they can resist the inflow of smoke.

prevention: the fire safety tactic of ensuring that fires do not start by controlling ignition and limiting fuel sources.

radiation: the transfer of heat without an intervening medium between the source and the receivers.

re-cycling (or re-setting) sprinkler system: a sprinkler system where the heads can be closed once the fire is extinguished such that water damage is minimized.

refuge: a place of safety within a building.

re-setting (or re-cycling) sprinkler system: a sprinkler system where the heads can be closed once the fire is extinguished such that water damage is minimized.

Glossary

rising mains: vertical pipes installed within tall buildings (usually over 18 m) which have a fire service connection or booster pump at the lower end and outlets at different levels within the building.

sacrificial timber: the technique of deliberately oversizing timber elements to enhance their fire resistance.

smoke: the general term for the solid and gaseous products of combustion in the rising plume of heated air, containing both burnt and unburnt parts of the fuel along with any gases given off by the chemical degradation of the fuel.

smoke curtains: barriers which can restrict the movement of smoke.

smoke layering (or smoke stratification): the process whereby different layers or zones develop within smoke due to the buoyancy of the gases involved.

smoke load: the amount of potential fuel within a building or room which will produce smoke.

smoke obscuration: reduction in visibility due to smoke.

smoke reservoirs: areas in the ceiling of a space in which smoke will collect.

smoke stratification (or smoke layering): the process whereby different layers or zones develop within smoke due to the buoyancy of the gases involved.

smoke venting: the technique of allowing smoke to escape from a building, or of forcing it out by mechanical means.

soot: fine particles, mainly carbon, produced and deposited during the incomplete combustion of organic materials.

spontaneous ignition temperature: the temperature to which a fuel has to be heated for the vapours given off to ignite without the application of an external ignition source.

sprinklers: auto-suppression systems using water sprays to extinguish small fires, and contain growing fires until the fire services arrive.

stage 1 escape: escape from the room or area of origin.

stage 2 escape: escape from the compartment or sub-compartment of origin by the circulation route to a final exit, a protected stair or an adjoining compartment offering refuge.

stage 3 escape: escape from the floor of origin to ground level.

stage 4 escape: final escape at ground level.

structural elements: loadbearing elements of the building, in particular floors and their supporting structures – e.g. columns or loadbearing walls.

structural protection: fire resistance provided to structural elements.

sub-compartments: fire and smoke tight areas into which a building can be divided to reduce travel distances, offering 30 min of resistance.

surface spread of flame: the potential for the spread of flame across the surface of a material.

toughened glass: glass which has been toughened to achieve higher levels of stability and integrity.

trade-offs: the technique of providing equivalent levels of fire safety through different fire safety measures.

traditional fire safety design: design which considers the building as a series of components and attempts to achieve fire safety by ensuring all such components meet a specified performance standard.

travel distance: the distance to be travelled from any point in a building to a place of safety, often specified in terms of the different stages of escape.

triangle of fire: a description of the three ingredients necessary for a fire (heat, fuel and oxygen).

wet sprinkler system: a sprinkler system where pipework is kept permanently charged with water.

wet rising main: a rising main normally kept permanently charged with water.

Index

Abrahams, John 88
Access, buildings, arson attack 35
Acrolein 14
Alcohol licences 129–30
Analogue addressable systems 50
Anderson, Jack 1
Applicants, building works 2–3, 5
Approved Document B 5, 88, 124–5, 131
Architects
  collaboration 6
  relations with fire officers 3
Architecture, user-orientated 57
Arson 32–3
  briefing, design and construction process 35–7
  defence 33–5
  risk 38–40
  types 37–8
Arson Prevention Bureau 32
ASET 61
Aspirating detectors 45–6
Association of British Insurers 32, 142
Association of Structural Fire Protection Contractors and Manufacturers 151
Atria, smoke control 110–19
Australian Building Fire Safety Systems Code 82

Beam detectors 51
BFIRES-II 61–2
Bickerdike Allen and Partners 1–2, 122
British Fire Protection Systems Association 152
British Gypsum Fire Research and Testing Laboratory 150

British Standards 19
  BS 476 18, 132–4
  BS 750 1984 134
  BS 1635 1990 135
  BS 3169 1986 135
  BS 4422 135
  BS 4547 1972 135
  BS 4569 1983 135
  BS 4790 1987 135
  BS 5041 135
  BS 5266 Part 1 1988 135
  BS 5268 135
  BS 5274 1985 135
  BS 5306 135–6
  BS 5378 136
  BS 5395 136
  BS 5423 1987 136
  BS 5445 136
  BS 5446 Part 1 1990 136
  BS 5499 136
  BS 5588 88, 136–7
  BS 5588 Part 8 63
  BS 5588 Part 10 80
  BS 5720 1979 137
  BS 5725 Part 1 1981 137
  BS 5839 137
  BS 5839 Part 1 47, 50–1, 53–4
  BS 5839 Part 2 46
  BS 5839 Part 4 50
  BS 5852 1990 137
  BS 5950 Part 8 1990 137
  BS 6266 1992 137
  BS 6336 1982 137
  BS 6459 Part 1 1984 137
  BS 6535 137–8
  BS 6575 1985 138
  BS 7175 1989 138

British Standards *cont'd*
  BS 7176 1991 138
  BS 7177 1991 138
  BS 7443 48
  BS 8110 138
  BS 8202 138
  PD 6496 1981 134
  PD 6512 137
  PD 6520 1988 134
Building Act 1984 3, 123
Building control authorities 1, 8
Building Control Officers 2–3
  collaboration 6
  education 3–4, 5
  negotiation with applicants 5
Building Regulations 3, 28–9, 31, 94, 99–100, 105, 107, 121, 123–6, 143, 144
  1985 139
  guidance 5–6
Building Regulations Advisory Committee 1, 6
Building Regulations (Northern Ireland) 1990 131
Building Research Establishment 139
Buildings
  arson attack 34–5
  arson risk 33
  existing fire safety 6–7
  high-rise 106, 107
    evacuation 61
  occupancy characteristics 89–91
  property protection 16–17
Building Standards (Scotland) Regulations 126–7
Burning, rate of 19, 106

Call points, manual 46–7
Carbon dioxide 19
Carbon monoxide 14, 15, 19
Ceilings, effects of fire 12
Cinemas 91
Cinematograph Act 1985 130
Circulation, routes within buildings 35, 39
Closed-circuit television 35
Clubs 91
Combustion 10–11
Communications, fire safety 22, 24, 25, 94, 95

Compartmentation 18, 92–3, 95, 97
  differing categories of building 99
  essential 103–5
  needs 101–3
  optional 105
Computer simulation
  design and 96
  escape behaviour 61–2
  evacuation 82
Concrete, reinforced, effects of fire 17
Cone calorimeter 19
Containment, fire safety 22, 25, 26
Control equipment, fire alarms 48–50
Coping, fire behaviour 62
Crowds
  evacuation research 60–1, 81
  safety management 57

Defend in place, fire escape behaviour 93
Department for Education 146
Department of the Environment 4, 5, 6, 122, 139
Department of Health, NHS Estates 145–6
Department of Trade and Industry, Enterprise and Deregulation Unit 1, 122
Design, against arson 35–7
Detection, automatic 18, 19
Detectors
  choice 41–3
  control equipment 48–50
  installation requirements 50–2
Disabled people, evacuation 63
Doors, furniture 36
Drysdale, Donald 9

Education, fire safety 4
EGRESS 82
  escape behaviour 92, 93
Environmental Design Evaluation 57–9
Escape
  design, first principles approach 94–6
  fire safety 22, 24–5, 26
Escape behaviour 56–9, 62
  department store fire 69–73
  exit choice 63–4

Escape behaviour *cont'd*
  factors 64–6
  lecture theatre evacuation 73–7, 81
  models 59–60
  nurses' hall of residence fire 67–9
  occupancy characteristics and 89–91
  research 60–3
    implications for design 79–82
  strategies 92–4
  underground station evacuation 77–9
  see also Life safety
European Community
  fire safety tests 18–19
  Framework Directive 89/391/EEC 24
  Workplace Directive 89/654/EEC 24
European Standards, EN54 44–5
EVACNET 61
Evacuation 88
  closed institutions 94–5
  computer simulation 82
  high-rise buildings 61
  lecture theatre 73–7
  times 79, 81
  underground station 77–9
Exhaust, smoke 119
EXIT89 61, 62
Exit choice, escape behaviour 63–4
EXITT 61
Extinguishing, fire 22, 25, 26, 94

Fault warning, wired systems 52
Fences, arson attack 34
Fire
  critical stages 9–10
  detection 41–6
  location indicators 52–3
  severity 17
  sizes 113–14
Fire alarms
  installation requirements 50–2
  sounders 47–8
  users' responsibilities 53–4
Fire authorities 1, 8
Fire Brigades 3, 42, 53
Fire cells 97
Fire Certificates 2, 127–8, 132
Fire Certificates (Special Premises) Regulations 1976 129
FIRECODE 145–6

Fire containment 92–3, 94, 97–8
  historical aspects 99–100
Fire engineering 4, 6, 30
  implications of escape behaviour research 79–82
Fire extinguishing, systems 106–7
Fire Extinguishing Trades Association 152
Fire-fighters, safety 100
Fire Grading Report 99
Fire Insurers' Research and Testing Organization 142
Fire loads 17, 30
Fire Officers, collaboration 6
Fire Offices' Committee 142, 143
Firepoint 11
Fire Precautions Act 1971 29, 94, 123, 127–8, 129
Fire prevention 9, 18, 22, 24, 25, 94
Fire prevention, arson 32–40
Fire Prevention Officers 2–3
  education 4, 5
Fire Protection Association 142, 143, 150
Fire Research Station 62–3, 111, 114, 139–42, 150
Fire resistance
  provision 107–8
  requirements 105–7
Fire safety 9, 22–3
  acceptability and equivalency 27–9
  application of fire science 18–19
  components 23–7
  'Design Against Fire' course 153–60
  education 3–4
  existing buildings 6–7
  glossary 161–6
  legislation 122–32
  life safety, *see* Life safety
  organizations 139–52
  procedure 2, 4–5
  property protection, *see* Property protection
  responsible person 53–4
  standards 132–8
  tactics 22–3
  tests 18–19
  traditional and holistic approaches 29–31

Fire Safety and Safety at Places of
 Sport Act 1987 128
Fire Safety Studies Working Group 4
Fire science 6, 9–17
 application to fire safety 18–19
Fire Service College 4
Fire Services Act 1947 3
Fire Services (Northern Ireland) Order
 1984 132
Fire signals 19
Fire zones 97–8
Flame
 radiation 41, 46
 spread 12, 18
Flame detectors 42–3, 46
Flashover 9
 prevention and delay 11–14, 16
Fraudulent arson 37
Fuel 106
 risk 25, 26
Fuel load, buildings 17

Gaming Act 1968 129
Gaming licences 129–30
Gases, smoky 112–13
Gasification, solids 10–11
Great Fire of London (1666) 99

Health and Safety at Work Act 1974
 123, 129
Health and Safety Executive/
 Inspectorate 129
Heat 21–2, 41
Heat detectors 42, 43–5
 positioning 51
Heat release, rate of 19
Heat transfer 11, 12
High-rise buildings 106, 107
 evacuation 61
Historic buildings, fire safety design
 95
Home Office 4, 62, 66, 79, 122, 127,
 130
 guidance on standards 127
 Working Group on the Prevention
  of Arson 32
Hotels 90
Human behaviour, modelling
 59–60
Hydrogen chloride 15

Hydrogen cyanide 14, 15

Ignition 9, 12, 18
 prevention 10–1
 risk 25, 26, 30
Infrared radiation 46
Institute of Fire Safety 151
Institution of Fire Engineers 151
Institutions, closed, means of escape
 94–5
Insulation 134
Insurance 29–30, 33, 36, 105
Insurance Technical Bureau 142
Integrity, structures 134
International Standards
 ISO 834 1975 134
 ISO 1182 1983 134
 ISO 1716 1973 134
 ISO 3008 1976 134
 ISO 3009 1976 134
 ISO 3261 1975 134
 ISO 5657 1986 133
 ISO 5925 134
 ISO 6944 1985 134
International Standards Organization
 44, 45
Intumescent Fire Seals Association 152
Ionization chamber smoke detectors
 45

Johnston, Lawrence 32
Joinery, second-fix 36

Keys, security 36–7
King's Cross fire 77, 78

Legislation 1, 27–8, 29
 British 122–30
 Northern Ireland 131–2
Licensing Act 1964 129
Life safety 9, 14–16, 21–2, 30, 99, 143
 compartmentation and 100–2, 104–5
 risk 25, 26
 see also Escape behaviour
Lighting, arson attack 34
Lighting systems, wayfinding 80
Line-type detectors 43–4
Lloyd's 142
Loadbearing capacity, structures 133
Local Acts 130–1

Index

Local authorities 130
London County Council 99
London District Surveyors' Association 144
London Fire and Civil Defence Authority 144
Loss Prevention Certification Board 142
Loss Prevention Council 142–4, 150

Maintenance
  long-term 38
  ongoing, arson defence 34–5
Malhotra, H. L. 97
Manual call points 46–7
Materials
  selection 18
  testing 18–19
Mobility, escape behaviour 90
Model fire scenarios 18
Morgan, Howard 110

National Bureau of Standards 61, 62
National Fire Protection Association 19, 62, 114
  Life Safety Code 61, 63
National Health Service, Crown Exemption 6–7
National Measurement Accreditation Service 150
National Supervisory Council for Intruder Alarms 142
NHS Estates 145–6
Northey, John 41

Optical beam detectors 42
Optical smoke detectors 45
Ove Arup 96

Part B Compliance Certificates 2, 5
People
  familiarity with buildings 90–1
  numbers, fire risk 89–90
Physical science, model of human reactions 59, 65, 80
Plumes
  entrainment 115
  spill 115–17, 118
Point detectors 42, 43, 44–5
  smoke 45

Polyurethane foam 15
Post-disaster Evaluation 57–9
Post-occupancy Evaluation 57–9
Production processes, arson risk 39
Property protection 9, 16–17, 21–2, 30, 99, 143
  compartmentation and 102–3, 105
  risk 25, 26
Psychological model, human reactions 59–60, 62, 65–6, 80
Public address systems 80

Recognition, fire behaviour 62
Refuge, fire escape behaviour 93, 94
Reinforced concrete, effects of fire 17
Repair, ongoing, arson defence 34–5
Rescue, means of escape 94
Risk 25, 26
  acceptability and equivalency 27–9
  arson 38–40
  assessment 30–1
RSET/ASET 62

Sampling detectors 45–6
Schools, arson attacks 33
Scottish Home and Health Department 127, 130
Search distance, fire location 52–3
Shopping malls, smoke control 110–19
Sime, Jonathan 56
Sites, perimeters, arson attack 34–5
Sleeping risk, fire response time 89
Smoke 18, 41
  damage to buildings 16
  logging 26, 114
  production 14–16, 19
Smoke control 18
Smoke detectors 42, 45
  positioning 51
Smoke ventilation 111–12
  basic principles 112–14
  design parameters 114–19
Social science, model of human reactions 59–60, 62, 65–6
Society of Fire Protection Engineers 151
Society of Fire Safety Engineers 151
Sounders 47–8
South-West Regional Health Authority 95

Spill plumes 115–16, 118
Sprinklers 15, 26, 28, 36, 106–7, 113, 114–15
Stansted Airport 96
Stardust Disco fire 90
Steel, effects of fire 16–17
Steel Construction Institute 147–8
Stollard, Paul 21, 121
Storage buildings 99
Structural damage, fire 16–17
Surveillance, passive, arson defence 35

Temperature, firepoint 10–11
Theatre Act 1968 130
Timber, effects of fire 17
Timber Research and Development Association 95, 146–7
Toxicity, smoke 18

Tyne and Wear Passenger Transport Executive 66, 77

Ultraviolet radiation 46
User-orientated architecture 57

Vandalism 37
VEGAS 82
Ventilation
  fire control 12, 14
  smoke, *see* Smoke ventilation
Voice alarms 48

Warehouses 99
  safe storage 38–9
Warning devices, fire 47–8
Warrington Fire Research Centre 150
Waste, storage, arson risk 39

Yarsley Ltd 150